THE PERIODIC TABLE

A Visual Guide to the Elements

TOM JACKSON

First published in Great Britain
2017 by Aurum Press an imprint of The Quarto Group

This edition first published in 2020 by White Lion Publishing,
an imprint of the Quarto Group
1 Triptych Place,
2nd Floor, London,
SE1 9SH,
United Kingdom
T (0) 20 7700 6700
www.Quarto.com

A catalogue record for this book is available from the British Library.
ISBN 978-1-78131-930-7
eBook ISBN 978-1-78131-931-4

7 9 10 8 6

Designed by The Urban Ant Ltd
Main illustrations by Andi Best
All images © Shutterstock.com

Printed in Malaysia

MIX
Paper | Supporting
responsible forestry
FSC™ C007207
FSC
www.fsc.org

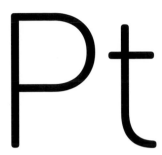

THE PERIODIC TABLE

A Visual Guide to the Elements

TOM JACKSON

WHITE LION
PUBLISHING

CONTENTS

The Periodic Table
1

The Big Picture
2

Inside Chemistry
3

4 Directory of Elements

4

INTRODUCTION

The periodic table is the ultimate infographic. It presents the fabric of the Universe (at least the bits of it we can see) as 118 units, and we can learn a lot about these units – the chemical elements – just by looking at where they sit among the rest.

The elements are substances that cannot be refined or purified into simpler ingredients. Each one is unique, with a set of physical and chemical properties that arises from the structure of its atoms. In 1869, Dmitri Mendeleev, a Russian chemist, created the periodic table as a way of organizing the known elements (around half of what is there today) into a system that related the increasing weights of each element with patterns in their chemical properties. Although he did not know it, Mendeleev's system was based around the varying atomic structures of the elements. He was working 30 years before the first subatomic particle, the electron, was discovered, and 60 years before researchers had drawn a full picture of how atoms are constructed from smaller particles. Nevertheless, that picture revealed why the periodic table works so well. Every element has a unique collection of subatomic particles –

electrons, protons and neutrons – and the way these particles are arranged gives an element its set of characteristics.

Inside the pages of this book you will find out how the particles do that, and see the great range of properties that elements have to offer. Despite following the same set of rules, the diversity of the elements is staggering. Some atoms have existed since the dawn of time – and will exist until the last moment of the Universe – while others fade away in a millisecond after being formed inside the crucible of a dying star (or inside an earthbound laboratory).

Not every element can offer up a superlative; most occupy a middle ground. However, that middle ground is the sum total of material in the Universe. It includes the metals that make our magnets, engines and electrical devices. It contains the semiconductors that have created the modern world through computing and look set to save the future with solar power. And it consists of the non-metal elements that fuel and maintain life on Earth, and, it is assumed, elsewhere, too. So let's begin a visual journey into the stuff of nature.

THE PERIODIC TABLE

THE PERIODIC TABLE

1 H HYDROGEN

1

The periodic table shows the elements in order of atomic number (the number of protons their atoms have). The elements are also arranged in rows, known as periods, so that elements with similar chemistry are lined up into columns, or groups. In this version, elements with similar properties are shown by the colour categories identified in the key on the right.

4 Be BERYLLIUM

12 Mg MAGNESIUM

20 Ca CALCIUM	21 Sc SCANDIUM	22 Ti TITANIUM	23 V VANADIUM	24 Cr CHROMIUM	25 Mn MANGANESE	26 Fe IRON	27 Co COBALT
38 Sr STRONTIUM	39 Y YTTRIUM	40 Zr ZIRCONIUM	41 Nb NIOBIUM	42 Mo MOLYBDENUM	43 Tc TECHNETIUM	44 Ru RUTHENIUM	45 Rh RHODIUM
56 Ba BARIUM	57–71 LANTHANIDES	72 Hf HAFNIUM	73 Ta TANTALUM	74 W TUNGSTEN	75 Re RHENIUM	76 Os OSMIUM	77 Ir IRIDIUM
88 Ra RADIUM	89–103 ACTINIDES	104 Rf RUTHERFORDIUM	105 Db DUBNIUM	106 Sg SEABORGIUM	107 Bh BOHRIUM	108 Hs HASSIUM	109 Mt MEITNERIUM

57 La LANTHANUM	58 Ce CERIUM	59 Pr PRASEODYMIUM	60 Nd NEODYMIUM	61 Pm PROMETHIUM	62 Sm SAMARIUM
89 Ac ACTINIUM	90 Th THORIUM	91 Pa PROTACTINIUM	92 U URANIUM	93 Np NEPTUNIUM	94 Pu PLUTONIUM

ALKALI METALS
A group of reactive metals occupying the far-left column of the periodic table. They are all soft, but they are solid metals at room temperature and never appear pure in nature.

ALKALINE EARTH METALS
The alkaline earth metals are silver-white metals at room temperature. The name is a term that refers to the many naturally occurring oxides of these elements found in, or derived from, rocks. For example, lime is the alkaline oxide of calcium.

LANTHANIDES
The lanthanide elements occupy a horizontal strip normally appended at the foot of the periodic table. Named after lanthanum, the first element in the series, they are generally found in large amounts in uncommon minerals such as monazite.

ACTINIDES
The actinides make up the second horizontal strip at the foot of the table. Named after their first element, actinium, they are all highly radioactive and include the main sources of nuclear fuels.

TRANSITION METALS
The transition metals make up the central block of the periodic table. They are harder than the alkali metals, less reactive, and are generally good conductors of both heat and electrical current.

POST-TRANSITION METALS
Also known as the poor metals, this triangular region contains unreactive metals that have weak metallic properties. They mostly have low melting and boiling points.

ELEMENT CATEGORIES

- Alkali metals
- Alkaline earth metals
- Lanthanides
- Actinides
- Transition metals
- Post-transition metals
- Metalloids
- Other non-metals
- Halogens
- Noble gases
- Unknown chemical properties

									2 He HELIUM
			5 B BORON	6 C CARBON	7 N NITROGEN	8 O OXYGEN	9 F FLUORINE		10 Ne NEON
			13 Al ALUMINIUM	14 Si SILICON	15 P PHOSPHORUS	16 S SULPHUR	17 Cl CHLORINE		18 Ar ARGON
28 Ni NICKEL	29 Cu COPPER	30 Zn ZINC	31 Ga GALLIUM	32 Ge GERMANIUM	33 As ARSENIC	34 Se SELENIUM	35 Br BROMINE		36 Kr KRYPTON
46 Pd PALLADIUM	47 Ag SILVER	48 Cd CADMIUM	49 In INDIUM	50 Sn TIN	51 Sb ANTIMONY	52 Te TELLURIUM	53 I IODINE		54 Xe XENON
78 Pt PLATINUM	79 Au GOLD	80 Hg MERCURY	81 Tl THALLIUM	82 Pb LEAD	83 Bi BISMUTH	84 Po POLONIUM	85 At ASTATINE		86 Rn RADON
110 Ds DARMSTADTIUM	111 Rg ROENTGENIUM	112 Cn COPERNICIUM	113 Nh NIHONIUM	114 Fl FLEROVIUM	115 Mc MOSCOVIUM	116 Lv LIVERMORIUM	117 Ts TENNESSINE		118 Og OGANESSON
63 Eu EUROPIUM	64 Gd GADOLINIUM	65 Tb TERBIUM	66 Dy DYSPROSIUM	67 Ho HOLMIUM	68 Er ERBIUM	69 Tm THULIUM	70 Yb YTTERBIUM		71 Lu LUTETIUM
95 Am AMERICIUM	96 Cm CURIUM	97 Bk BERKELIUM	98 Cf CALIFORNIUM	99 Es EINSTEINIUM	100 Fm FERMIUM	101 Md MENDELEVIUM	102 No NOBELIUM		103 Lr LAWRENCIUM

METALLOIDS

The metalloid elements form a divide between the metals and non-metals in the periodic table. Their electrical properties are intermediate between these other two categories, leading to their use in semiconductor electronics.

OTHER NON-METALS

A loose collection of elements that do not fall into the halogens and noble gases and are therefore shown here as a separate group. However, they display a wide range of chemical and physical properties. Most non-metals have the ability to gain electrons easily. They generally have lower melting points, boiling points and densities than the metal elements.

HALOGENS

The halogens, known as group 17, are the only group to contain all three principal states of matter at room temperature: gas (fluorine and chlorine), liquid (bromine) and solid (iodine and astatine) – all non-metals.

NOBLE GASES

The noble gases are non-metals occupying Group 18 of the table. They are all gaseous at room temperature and share the properties of being colourless, odourless and unreactive. Including neon, argon and xenon, they have applications in lighting and welding.

UNKNOWN CHEMICAL PROPERTIES

Elements larger than uranium are generally made in the laboratory, and very often only in minute quantities. The chemical properties of several of the latest and largest artificial elements remain something of a mystery.

ATOMIC STRUCTURE

1

The idea of the atom sounds like something very modern. After all, scientists are still trying to solve some of its mysteries. However, the concept of atoms was considered by ancient philosophers 2,500 years ago, and has been central to our understanding of chemistry for at least 200 years.

ELEMENTS

Ancient cultures understood nature in terms of elements, or the basic materials that made everything on Earth. The most common set contained four elements: earth, water, air and fire.

NATURAL PROCESSES

Aristotle, the Greek thinker, thought that the ever-changing nature of the Universe was due to the elements trying to separate from each other.

SIMPLE PROPERTIES

The four elements were thought to give every substance its basic properties, making them feel cold, hot, dry or wet.

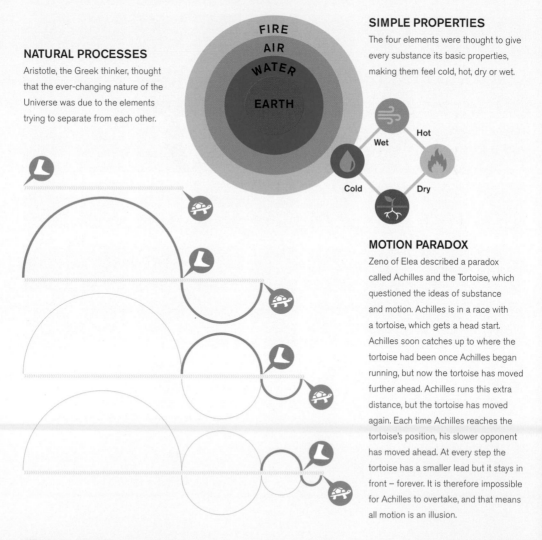

FIRE
AIR
WATER
EARTH

Wet · Hot
Cold · Dry

MOTION PARADOX

Zeno of Elea described a paradox called Achilles and the Tortoise, which questioned the ideas of substance and motion. Achilles is in a race with a tortoise, which gets a head start. Achilles soon catches up to where the tortoise had been once Achilles began running, but now the tortoise has moved further ahead. Achilles runs this extra distance, but the tortoise has moved again. Each time Achilles reaches the tortoise's position, his slower opponent has moved ahead. At every step the tortoise has a smaller lead but it stays in front – forever. It is therefore impossible for Achilles to overtake, and that means all motion is an illusion.

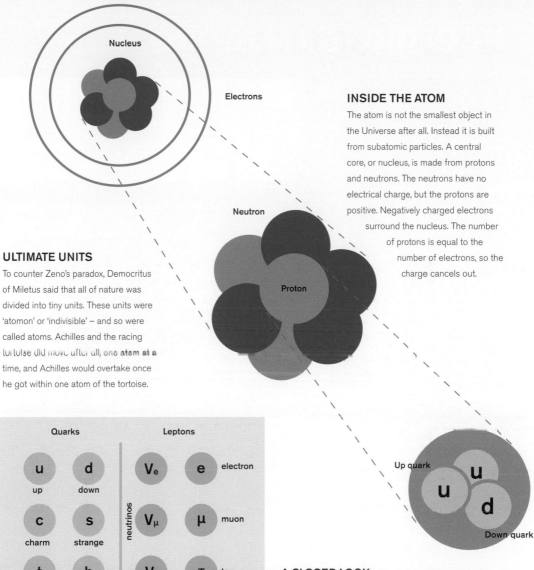

Nucleus

Electrons

Neutron

Proton

INSIDE THE ATOM

The atom is not the smallest object in the Universe after all. Instead it is built from subatomic particles. A central core, or nucleus, is made from protons and neutrons. The neutrons have no electrical charge, but the protons are positive. Negatively charged electrons surround the nucleus. The number of protons is equal to the number of electrons, so the charge cancels out.

ULTIMATE UNITS

To counter Zeno's paradox, Democritus of Miletus said that all of nature was divided into tiny units. These units were 'atomon' or 'indivisible' – and so were called atoms. Achilles and the racing tortoise did move after all, one atom at a time, and Achilles would overtake once he got within one atom of the tortoise.

Up quark

u

u

d

Down quark

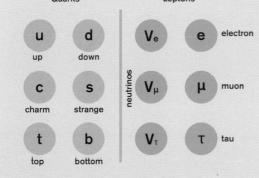

Quarks

u	d
up	down
c	s
charm	strange
t	b
top	bottom

Leptons

neutrinos

V_e	e	electron
V_μ	μ	muon
V_τ	τ	tau

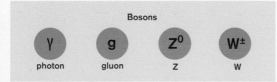

Bosons

γ	g	Z^0	$W^\pm$
photon	gluon	Z	W

A CLOSER LOOK

The three kinds of subatomic particles in an atom are not the end of the story. The Standard Model says that the Universe is built using 16 subatomic particles. A proton is made from three quarks, two ups and one down; a neutron is two downs and one up. Objects with mass are always collections of quarks and leptons (the electron is the main kind). The forces that control how objects behave are carried by particles called bosons.

HOW BIG IS AN ATOM?

The atom is the smallest unit of an element, and its size is hard to imagine on a human scale. To confound the issue, the particles that make up the atom cluster together, which means most of the space inside a tiny atom is empty anyway.

REAL-WORLD RATIO

The strongest microscope – the scanning tunnelling microscope – can detect the region occupied by a single atom. However, that is used to analyze the structure of materials. The images of atoms the microscope makes just show blobs, and we are still left struggling to picture their true size. The only way to imagine the size of an atom is to compare it with real-world objects. In this example, we use will use a penny coin and the moon.

Width of nucleus	Width of atom
Pea	Stadium
Beachball	Marathon course
London Eye	Pluto
Earth	Saturn's orbit

PENNY

A small coin like a penny is 170 million times wider than an atom.

ATOM

A hydrogen atom is about ten billionth of a metre wide.

THIS IS THE POINT

There are 7.5 trillion carbon and hydrogen atoms (mostly) in the full stop at the end of this sentence. That's roughly 1,000 for every person on Earth.

MOON

Just as a penny is to an atom, the moon is to a penny. In other words, it is about 170 million times wider. A penny dropped on the moon is like an atom resting on a penny.

EMPTY SPACE

Atoms are obviously tiny, but even so the particles inside take up very little space. The nucleus at the heart of the atom is 10,000 times smaller than the atom itself – and that is where nearly all the atom's matter is packed. The table (above left) helps with visualizing the size of a nucleus and size of the cloud of electrons that surrounds it.

99.9999999999996 per cent

of an atom – and by extension, every object in the Universe – is made up of nothing at all.

HOW THE TABLE WORKS

Instead of four elements, we now know that there are more than 100 of them – although only about 90 are naturally occurring. All elements are made from atoms. Every element has a unique number of protons, and this is its atomic number.

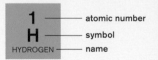

1
H
HYDROGEN

— atomic number
— symbol
— name

ELECTRON NUMBER

Atoms do not have an electric charge – they are always neutral. This is because the number of electrons in an atom is always equal to the atomic number.

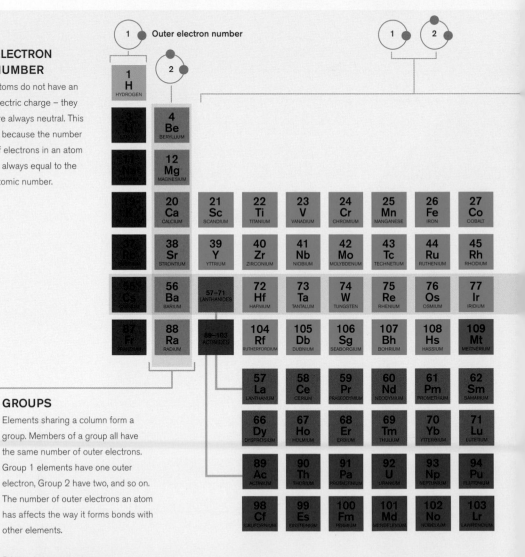

Outer electron number

GROUPS

Elements sharing a column form a group. Members of a group all have the same number of outer electrons. Group 1 elements have one outer electron, Group 2 have two, and so on. The number of outer electrons an atom has affects the way it forms bonds with other elements.

ELECTRON SHELLS

The electrons are arranged in shells around the nucleus. Each shell can hold a specific number of electrons.

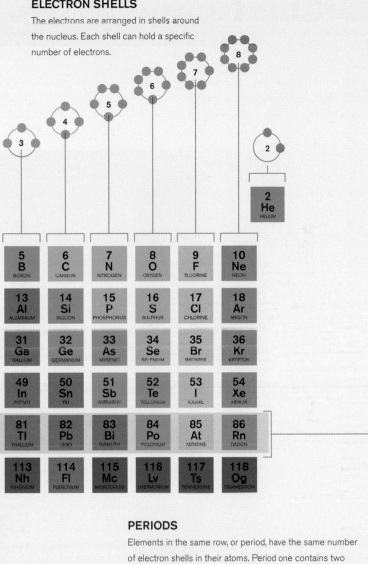

OUTER ELECTRONS

In most atoms, the outer electron shell is not full. The number of electrons in the outer shell gives the atom its properties.

2 He HELIUM		

5 B BORON	6 C CARBON	7 N NITROGEN	8 O OXYGEN	9 F FLUORINE	10 Ne NEON
13 Al ALUMINIUM	14 Si SILICON	15 P PHOSPHORUS	16 S SULPHUR	17 Cl CHLORINE	18 Ar ARGON

28 Ni NICKEL	29 Cu COPPER	30 Zn ZINC	31 Ga GALLIUM	32 Ge GERMANIUM	33 As ARSENIC	34 Se SELENIUM	35 Br BROMINE	36 Kr KRYPTON
46 Pd PALLADIUM	47 Ag SILVER	48 Cd CADMIUM	49 In INDIUM	50 Sn TIN	51 Sb ANTIMONY	52 Te TELLURIUM	53 I IODINE	54 Xe XENON
78 Pt PLATINUM	79 Au GOLD	80 Hg MERCURY	81 Tl THALLIUM	82 Pb LEAD	83 Bi BISMUTH	84 Po POLONIUM	85 At ASTATINE	86 Rn RADON
110 Ds DARMSTADTIUM	111 Rg ROENTGENIUM	112 Cn COPERNICIUM	113 Nh NIHONIUM	114 Fl FLEROVIUM	115 Mc MOSCOVIUM	116 Lv LIVERMORIUM	117 Ts TENNESSINE	118 Og OGANESSON

63 Eu EUROPIUM	64 Gd GADOLINIUM	65 Tb TERBIUM

95 Am AMERICIUM	96 Cm CURIUM	97 Bk BERKELIUM

PERIODS

Elements in the same row, or period, have the same number of electron shells in their atoms. Period one contains two elements, because the first electron shell has room for two electrons. Period two contains eight elements, because the second electron shell holds eight electrons. The third shell can hold 18 electrons, but only the first eight spaces fill up to start with. The remaining ones only fill up once the first two have entered the fourth shell. This creates the central sections, or series, of elements.

GROUP 1

Group 1 is also called the alkali metals. It includes sodium, potassium and other reactive metals. The metallic Group 1 elements react violently with water and can catch fire when exposed to the air. They are stored in oil to prevent explosions.

1

WHY NO HYDROGEN?

Hydrogen, which is a gas, not a metal, is technically a member of Group 1. However, as a gas that is made up of the lightest and simplest atom, hydrogen is treated as a special element on its own, with its own distinct chemistry.

CLOSER LOOK

• The name 'alkali metals' is derived from a universal property of the metallic members of the group. They all react with water to make a strongly alkaline compound. Alkalis are chemicals that react with acids to make a neutral compound known as a salt.

• Group 1 metals are all shiny when pure but will rapidly lose that lustre as they react with the air. They are also soft enough to be cut with a knife.

• The first three members of the group, lithium, sodium and potassium, have a density lower than water, so they float on the surface. The other three members sink.

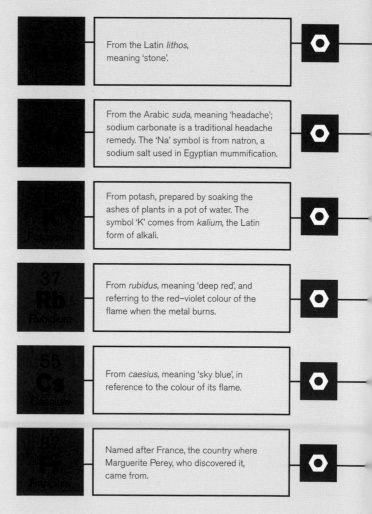

From the Latin *lithos*, meaning 'stone'.

From the Arabic *suda*, meaning 'headache'; sodium carbonate is a traditional headache remedy. The 'Na' symbol is from natron, a sodium salt used in Egyptian mummification.

From potash, prepared by soaking the ashes of plants in a pot of water. The symbol 'K' comes from *kalium*, the Latin form of alkali.

From *rubidus*, meaning 'deep red', and referring to the red–violet colour of the flame when the metal burns.

From *caesius*, meaning 'sky blue', in reference to the colour of its flame.

Named after France, the country where Marguerite Perey, who discovered it, came from.

OXIDATION STATE

All alkali metals form a single oxidation state of +1. This means that when they react, they lose the single outer electron to form an ion with a charge of +1.

MELTING POINTS

All alkali metals have low melting points. Caesium and francium would become liquid on a warm day.

Melting points

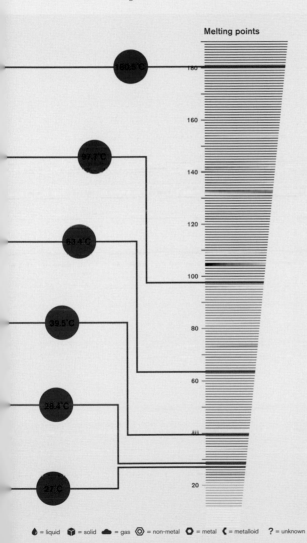

FLAME COLOURS

The alkali metals each produce flames of a distinctive colour as they burn. The same colours are produced when the elements are electrified as a gas. The orange light from sodium gas is used in some lighting.

Li   Na   K   Rb   Cs

● = liquid   ■ = solid   ☁ = gas   ◎ = non-metal   ⬡ = metal   〈 = metalloid   ? = unknown

GROUP 2

This group of metals is also known as the alkaline earth metals, because their 'earths', or powdery oxides, are always alkaline. All members have atoms with two outer electrons. Most Group 2 metals react with cold water to form hydroxide compounds and hydrogen gas. Only beryllium bucks the trend and is unchanged in water.

USEFUL METALS

Group 2 metals share a common atomic structure but that does not stop them being used in a wide variety of ways:

• Beryllium: When pure, this metal is transparent to X-rays – although light does not shine through, obviously. X-ray machines send out their scanning beam through a beryllium window.

• Magnesium: Milk of magnesia, a mixture of water and magnesium hydroxide, is a traditional cure for indigestion and constipation.

• Calcium: This metal is combined with phosphates to make the stiff material in bones and teeth.

• Strontium: This metal provides the red colour in fireworks.

• Barium: When combined with sulphate, this is swallowed to make the soft intestines show up in medical X-rays.

• Radium: Used as a source of radioactivity, and once thought to boost health but now heavily controlled.

| 4 Be Beryllium | From beryl, a pale gemstone that includes emerald, aquamarine and heliodor. |

| 12 Mg Magnesium | From Magnesia, a mineral-rich region in the north of Greece. The word 'magnet' also comes from this term, although magnesium is not a magnetic metal. |

| 20 Ca Calcium | From *calx*, the Latin word for 'lime', a caustic mineral made by heating chalk or limestone and used in cement, among many other things. |

| 38 Sr Strontium | Named after the Scottish village of Strontian, an area of lead mines where the first strontium minerals were identified. |

| 56 Ba Barium | From the mineral baryte, which is derived from the Greek word for 'heavy'. |

| 88 Ra Radium | From *radius*, the Latin word for 'ray', which is in reference to this metal's radioactivity. |

OXIDATION STATE

When they react, alkali metals lose their two outer electrons to form an ion with a charge of +2. Most of them form compounds in this way, although beryllium also forms covalent bonds. This is where their electrons share electrons with another atom.

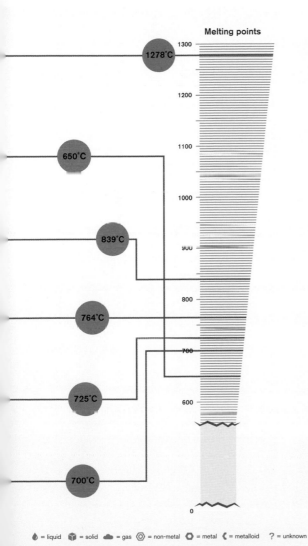

Melting points

1300
1278°C
1200
1100
650°C
1000
839°C
900
800
764°C
700
725°C
600
700°C
0

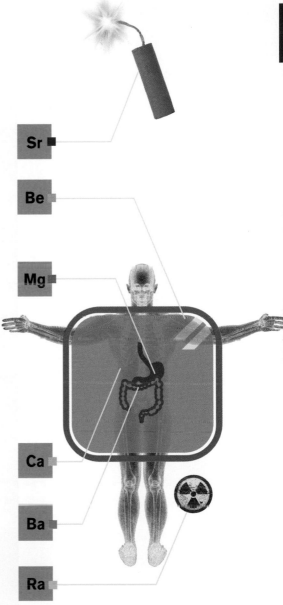

Sr ■

Be ▪

Mg ▪

Ca ▪

Ba ▪

Ra ▪

🌢 = liquid   📦 = solid   ☁ = gas   ◎ = non-metal   ⬡ = metal   ⟨ = metalloid   ? = unknown

GROUP 3

1

Also called the boron group for its first member, this set of elements is also described as 'triels', because they can bond to a maximum of three other atoms. However, only the lighter ones do this; heavier members tend to bond to one other atom at a time. Boron is one of the hardest elements, but the metallic members are all quite soft.

HEALTH EFFECTS – GOOD AND BAD

Members of Group 3 have an impact on human health, and not always in a good way.

• Boron: Although only needed in tiny amounts, this element is an essential nutrient in foods. It is used to help maintain strong bones.

• Aluminium: This is non-toxic, and has no role in the body. Claims that aluminium is linked to dementia and cancer are now thought to be wrong.

• Gallium: This metal is used as a last line of defence against the worst and most drug-resistant forms of malaria.

• Indium: Large amounts of this metal will damage the kidneys. Metalworkers are most often exposed to this element.

• Thallium: Tiny amounts of this metal cause vomiting and diarrhoea and as little as 15 milligrams will kill. It is reported that the CIA planned to poison Fidel Castro in 1959 by putting thallium salt, then used as a depilator, in his shoes. The idea – never carried out – was to make his iconic beard fall out.

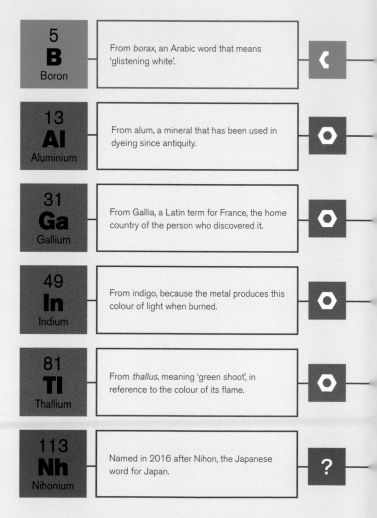

5 **B** Boron	From *borax*, an Arabic word that means 'glistening white'.
13 **Al** Aluminium	From alum, a mineral that has been used in dyeing since antiquity.
31 **Ga** Gallium	From Gallia, a Latin term for France, the home country of the person who discovered it.
49 **In** Indium	From indigo, because the metal produces this colour of light when burned.
81 **Tl** Thallium	From *thallus*, meaning 'green shoot', in reference to the colour of its flame.
113 **Nh** Nihonium	Named in 2016 after Nihon, the Japanese word for Japan.

OXIDATION STATES

All members of the group can form +3 ions by losing their outer electrons, although heavier members tend to produce more stable +1 ions.

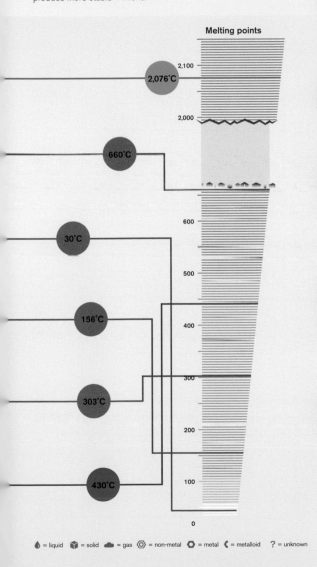

Melting points

2,076°C — 2,100

2,000

660°C

600

30°C

500

156°C — 400

303°C — 300

200

430°C — 100

0

🜄 = liquid   📦 = solid   ☁ = gas   ◎ = non-metal   ⬭ = metal   🌙 = metalloid   ? = unknown

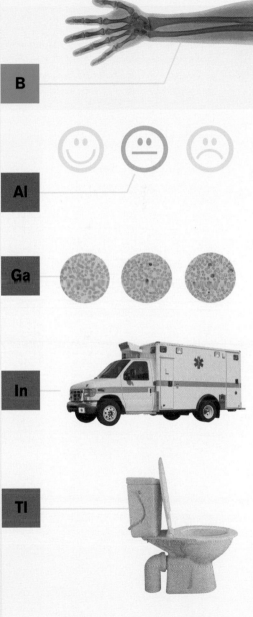

B

Al

Ga

In

Tl

GROUP 4

This group is also called the crystallogens because its members form the widest variety of crystals of any group in the periodic table. The four outer electrons in their atoms mean that these elements have outer shells that are half full. As a result, these elements can lose or gain electrons and form bonds with up to four other elements at a time.

INSIDE GROUP 4

• This group is the only one where all the members are solid in standard conditions, but also made up of non-metals, metalloids and metals.

• All members of Group 4 have pure forms that conduct electricity. For example, graphite carbon is highly conductive (but diamond is not). Silicon and germanium are semiconductors, which means they act as electrical insulators as well as conductors.

• The atoms of members can make a maximum of four bonds, sometimes forming double and even triple bonds with each other.

• Carbon and silicon are able to form chained, branched and ringed molecules. Out of ten million known compounds, 90 per cent contain carbon.

• Tin and lead can form +2 ions and thus have metallic properties.

• Silicate ions ($SiO4^{4-}$) are in 90 per cent of all rock-forming minerals found in the Earth's crust.

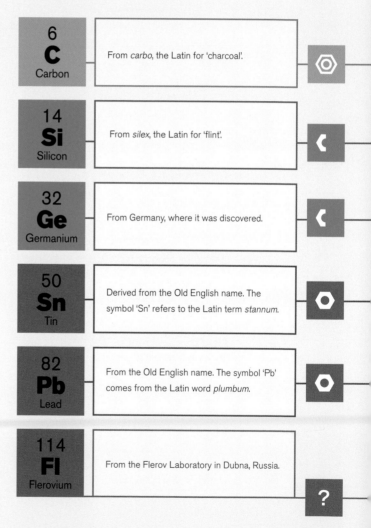

6 C Carbon	From *carbo*, the Latin for 'charcoal'.
14 Si Silicon	From *silex*, the Latin for 'flint'.
32 Ge Germanium	From Germany, where it was discovered.
50 Sn Tin	Derived from the Old English name. The symbol 'Sn' refers to the Latin term *stannum*.
82 Pb Lead	From the Old English name. The symbol 'Pb' comes from the Latin word *plumbum*.
114 Fl Flerovium	From the Flerov Laboratory in Dubna, Russia.

OXIDATION STATES

Most crystallogens lose outer electrons to form +4 ions. Only carbon is able to gain electrons to form a −4 ion, known as a carbide.

MELTING POINT

Carbon has the highest melting point of all elements, although it actually sublimates straight into a gas.

VARIED VALUE

The price of pure samples of Group 4 reflects their abundance in the Earth's crust and the cost of refining.

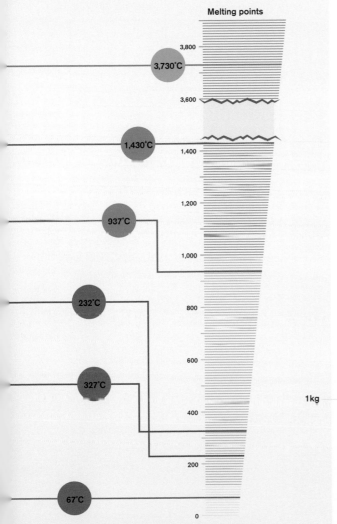

Melting points

3,800

3,730°C

3,600

1,430°C

1,400

1,200

937°C

1,000

232°C

800

327°C

600

1kg

400

200

67°C

0

$500,000

Heavy elements are generally rarer than lighter ones. Lead is an exception, being more abundant than tin because it is produced by the decay of many radioactive elements.

$1,000

$20

$0.50

$2

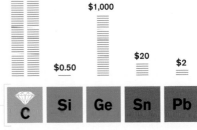

The diamond form of carbon is significantly more expensive because these crystals form deep in the Earth's mantle, and are only occasionally pushed up into the crust by volcanic activity.

💧 = liquid 🎲 = solid ☁ = gas Ⓝ = non-metal ⬭ = metal ◖ = metalloid ? = unknown

GROUP 5

1

Another diverse group of non-metals, metalloids and metals, this set of elements is also called the pnictogens. This means the 'choke makers', and refers to nitrogen, the first member, which makes up the portion of air that does not support life.

INSIDE GROUP 5

All members of this group can kill. Nitrogen is not toxic per se, since we breathe it in all the time. However, a pure nitrogen atmosphere results in asphyxiation. The other members can cause ill health and death.

Compounds of these elements have been used for thousands of years:

• Sal ammoniac is a nitrogen-rich mineral used in medicines and for dyeing in ancient Egypt. Saltpetre, also known as nitre, is a component in gunpowder.

• Calcium phosphate from bone ash was used to strengthen earthenware, making bone china. When first isolated, pure phosphorus was thought to be the philosopher's stone by alchemists.

• Arsenic comes from the mineral orpiment, used in golden paint in Rennaisance art and to make poison-tipped arrows.

• Powdered stibnite (antimony sulphide) was used as kohl, the dark eye make-up used in Egypt and Persia.

• Bismuth was used in Inca knives.

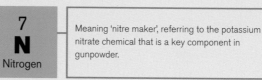

| 7 **N** Nitrogen | Meaning 'nitre maker', referring to the potassium nitrate chemical that is a key component in gunpowder. |

| 15 **P** Phosphorus | Named after the Greek term for the Morning Star, which means 'light giver'. Pure phosphorus can glow in certain conditions. |

| 33 **As** Arsenic | From the Arabic term *al zarniqa*, which means 'golden coloured'. This is in reference to the arsenic mineral orpiment, which is vibrant yellow. |

| 51 **Sb** Antimony | 'Antimony' means 'monk killer'; antimony poisoning claimed many early scientists who were often monks. The symbol 'Sb' comes from *stibium*, the Latin name for antimony. |

| 83 **Bi** Bismuth | From the Old German term *Wismuth*, which means 'white mass'. This is in reference to the pale mineral bismite. |

| 115 **Mc** Moscovium | Named after Moscow, Russia, the nearest city to the Joint Institute for Nuclear Research where this element was first synthesized. |

OXIDATION STATES

The upper members from nitrogen to arsenic produce mostly negative −3 ions. The lower members, antimony and bismuth, also form +3 and +5 ions.

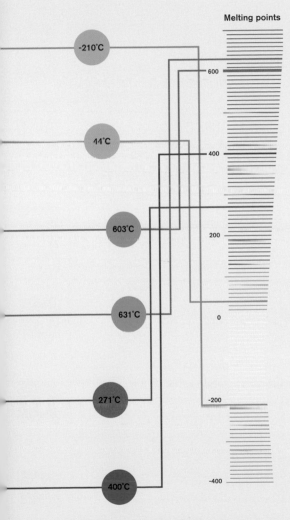

Melting points

-210°C

14°C

603°C

631°C

271°C

400°C

600

400

200

0

-200

-400

⬥ = liquid   ▣ = solid   ☁ = gas   ◎ = non-metal   ◯ = metal   ❰ = metalloid   ? = unknown

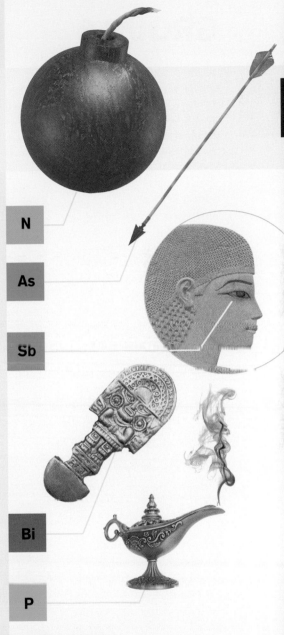

N

As

Sb

Bi

P

MAGIC MATERIAL

When pure phosphorus was first isolated in the 1660s, it was belived to be the philosopher's stone, a magical chemical that could make pure gold.

GROUP 6

Also known as the chalcogens, meaning the 'ore-formers', this group is dominated by oxygen and sulphur, two of the most common and reactive non-metals in the periodic table. Most of the ores for iron and other useful metals contain these elements. Other members are considerably rarer and have more niche applications.

1

INSIDE GROUP 6

Simple ionic compounds formed by Group 6 elements end in '-ide', such as oxide, sulphide, etc. Lower members form polyatomic ions with oxygen. With three oxygens, the ions end in '-ite', such as sulphite, etc. With four oxygens, they end in '-ate', such as sulphate, etc.

All Group 6 elements exist in several pure forms, or allotropes:

• Oxygen, four forms: dioxygen (O_2) is the gas in the air. It is pale blue when liquid; ozone (O_3) is a darker blue; oxozone (O_4) forms only when oxygen is liquefied; and O_8 molecules form when it is frozen, forming red oxygen.

• Sulphur, three forms: rhombic sulphur is yellow; monoclinic sulphur is orange; and plastic sulphur is black.

• Selenium, three forms: black, grey and red selenium.

• Tellurium, two forms: crystalline tellurium is metallic silver; amorphous tellurium is brown and powdery.

• Polonium, two forms: cubic and rhombohedral crystals.

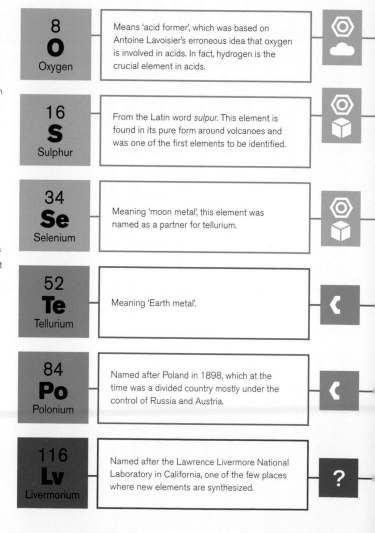

8
O
Oxygen

Means 'acid former', which was based on Antoine Lavoisier's erroneous idea that oxygen is involved in acids. In fact, hydrogen is the crucial element in acids.

16
S
Sulphur

From the Latin word *sulpur*. This element is found in its pure form around volcanoes and was one of the first elements to be identified.

34
Se
Selenium

Meaning 'moon metal', this element was named as a partner for tellurium.

52
Te
Tellurium

Meaning 'Earth metal'.

84
Po
Polonium

Named after Poland in 1898, which at the time was a divided country mostly under the control of Russia and Austria.

116
Lv
Livermorium

Named after the Lawrence Livermore National Laboratory in California, one of the few places where new elements are synthesized.

OXIDATION STATES

All members can form −2 ions by gaining two electrons. Polonium can lose electrons to form +2 and +4 ions and so is very metallic. By combining with oxygen, these elements form polyatomic ions where the oxidation state of the non-oxygen atom is +6.

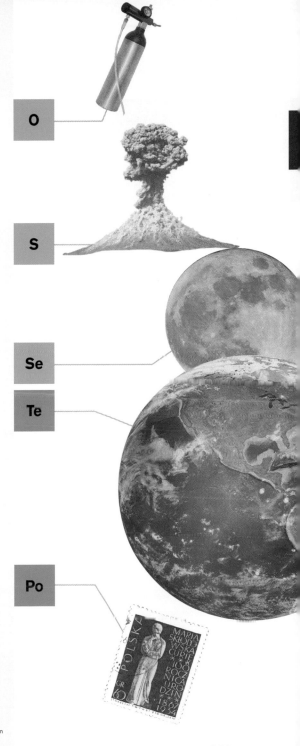

O

S

Se

Te

Po

Melting points

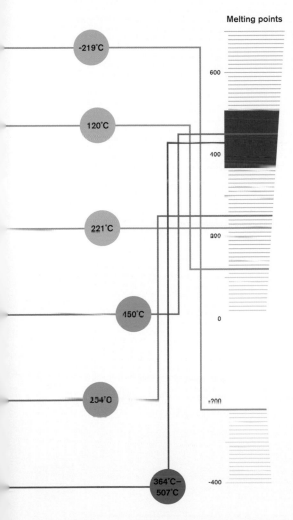

-219°C

600

120°C

400

221°C

200

450°C

0

254°O

-200

364°C–
507°C

-400

💧 = liquid 🧊 = solid ☁️ = gas ◎ = non-metal ⬡ = metal ☾ = metalloid ? = unknown

GROUP 7

Also known as the halogens, this group of non-metals includes some of the most reactive elements of the lot, most notably fluorine. The term halogen means 'salt-former' because halogens form stable, solid compounds known as salts, with names that end in '-ide'. The most familiar is sodium chloride, better known as common salt.

USES OF HALOGENS

All halogens are toxic in large enough quantities, but many are also used in the body or to promote health, hygiene and wellbeing.

• Fluorine: While pure fluorine is highly reactive and damaging, fluoride salts are used in toothpaste to strengthen the chemicals in tooth enamel.

• Chlorine: Bleach and other cleaning products are based on the chemical action of chlorine. The chlorine also attacks coloured compounds, stopping them absorbing light so they always appear white.

• Bromine: Bromine compounds are used as fire retardants. Heat from a fire releases pure bromine atoms, which interfere with the combustion process.

• Iodine: Iodine is a mild but powerful antibacterial agent used to sterilize wounds. As the salt silver iodide, it is the active ingredient in photographic film.

• Astatine: This radioactive halogen must be synthesized in a laboratory. So far it has not been put to any uses.

9 **F** Fluorine	Named after the mineral fluorite, which itself is named for its use as a flux, or a material that is used to remove impurities during metal refining.
17 **Cl** Chlorine	From *khlôros*, the ancient Greek term for 'green'. Pure chlorine is a pale-green gas.
35 **Br** Bromine	Named after the Greek word for 'stench', in reference to the strong, sharp smell of bromine vapour.
53 **I** Iodine	From the Greek word for 'violet', referring to the colour of the vapour that sublimates from solid iodine.
85 **At** Astatine	Named using the Greek word *astatos*, which means 'unstable'. Astatine is so radioactive that only a few grams exist on Earth at a time.
117 **Ts** Tennessine	Named for Tennessee, the home of the Oak Ridge National Laboratory, which was one of the labs involved in the synthesis of this element.

OXIDATION STATES

Halogens most commonly take an oxidation state of −1, meaning that when they react, they gain one outer electron to form an ion with a charge of −1.

Melting points

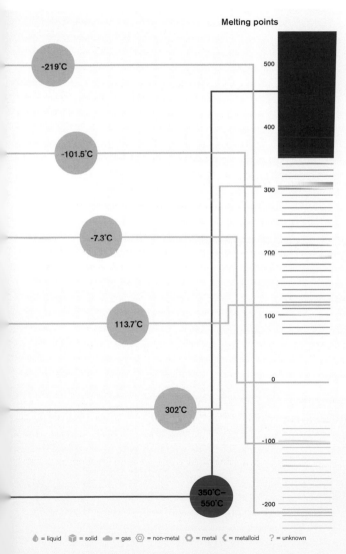

F

-219°C

CL

-101.5°C

-7.3°C

Br

113.7°C

I

302°C

At

350°C–550°C

500

400

300

200

100

0

-100

-200

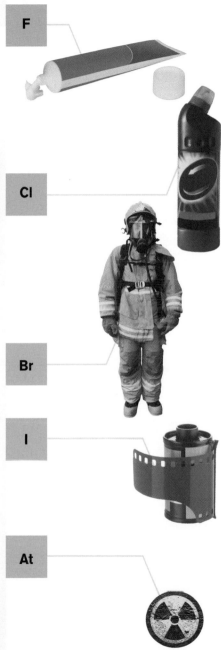

💧 = liquid   🧊 = solid   ☁ = gas   ◎ = non-metal   ○ = metal   ℂ = metalloid   ? = unknown

GROUP 8

1

Some chemists prefer to call this Group 0, because its members have atoms with eight outer electrons. This gives them a full outer shell and thus zero electrons available to take part in reactions. Group 8 elements are inert; they do not really react. As such they are called the noble gases – they don't mix with the common elements.

THE NOBLE GASES

All members of Group 8 are gases – at least as far as we know. A few atoms of an artificial member, known as oganesson, have recently been synthesized, but it is too early to describe its physical properties.

• While other gaseous elements such as hydrogen, flourine and nitrogen all contain diatomic molecules, such as H_2, N_2 or F_2, where two atoms are bonded together, the noble gases are simply clouds of single atoms. Noble gas atoms cannot form bonds with each other.

• The density of the noble gases increases down the group. Helium is the second-lightest element after hydrogen. Neon is lighter than air. Argon and krypton are slightly heavier, while xenon and radon are considerably denser than air, and they form the closest thing we can get to the proverbial 'lead balloon'.

• In laboratory conditions it has been possible to make krypton, xenon and radon bond with fluorine by forcing the gases to release an electron from deep inside their atoms to make a +1 ion.

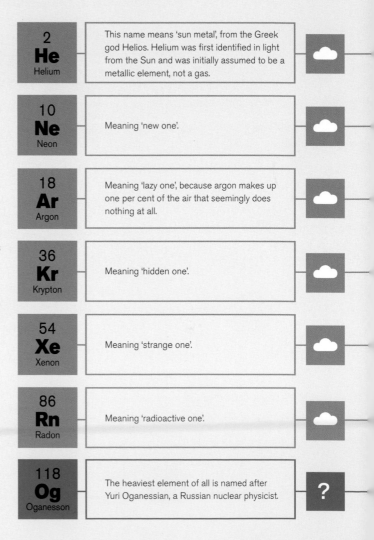

2 **He** Helium	This name means 'sun metal', from the Greek god Helios. Helium was first identified in light from the Sun and was initially assumed to be a metallic element, not a gas.
10 **Ne** Neon	Meaning 'new one'.
18 **Ar** Argon	Meaning 'lazy one', because argon makes up one per cent of the air that seemingly does nothing at all.
36 **Kr** Krypton	Meaning 'hidden one'.
54 **Xe** Xenon	Meaning 'strange one'.
86 **Rn** Radon	Meaning 'radioactive one'.
118 **Og** Oganesson	The heaviest element of all is named after Yuri Oganessian, a Russian nuclear physicist.

OXIDATION STATES

The noble gases do not lose or gain electrons because they already have a full outer shell of electrons in their atoms. Their oxidation state remains as 0.

GAS LIGHT

All noble gases glow with a distinctive colour when electrified. This is the basis of what is known as 'neon lighting', because neon was the first gas to be used like this.

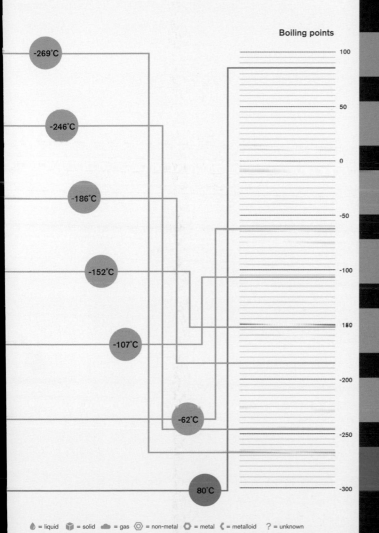

Boiling points

-269°C

-246°C

-186°C

-152°C

-107°C

-62°C

80°C

100

50

0

-50

-100

150

-200

-250

-300

He
Ne
Ar
Kr
Xe
Rn
Og

💧 = liquid   🎲 = solid   ☁ = gas   ◎ = non-metal   ⬡ = metal   ❲ = metalloid   ? = unknown

THE TRANSITION SERIES

1

Once we reach the fourth period of the periodic table, the concept of groups of elements breaks down. After Group 2, the atomic number continues to rise but the number of outer electrons stays the same. This forms a block of metallic elements – including many familiar ones – that is known as the transition series.

INFILLING ELECTRONS

Members of the transition series have atoms that follow the same rules as the rest of the elements: the number of electrons always equals that of the protons. However, additional electrons are not added to the outside of the atom. Instead, they fill spaces in the next electron shell down.

METALS ALL

All 38 elements in the transition series are metals. This is due to the number of outer electrons. Most have two outer electrons, but 12 elements, including copper, silver and gold, have just a single outer electron.

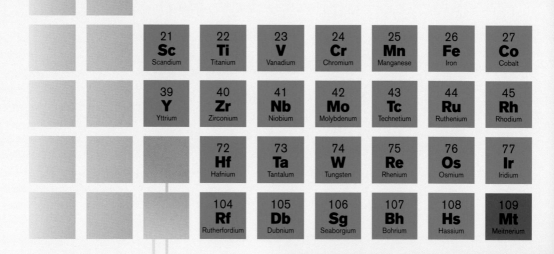

21 Sc Scandium	22 Ti Titanium	23 V Vanadium	24 Cr Chromium	25 Mn Manganese	26 Fe Iron	27 Co Cobalt
39 Y Yttrium	40 Zr Zirconium	41 Nb Niobium	42 Mo Molybdenum	43 Tc Technetium	44 Ru Ruthenium	45 Rh Rhodium
	72 Hf Hafnium	73 Ta Tantalum	74 W Tungsten	75 Re Rhenium	76 Os Osmium	77 Ir Iridium
	104 Rf Rutherfordium	105 Db Dubnium	106 Sg Seaborgium	107 Bh Bohrium	108 Hs Hassium	109 Mt Meitnerium

SAME AGAIN

The fifth shell will do something similar to the fourth. Once we reach the sixth period, the fifth shell will start to infill, and this time it has 24 empty spaces. This creates the inner transition series, more commonly known as the lanthanides and actinides.

INNER SHELLS

The first electron shell in an atom has room for two electrons – and the first period has two elements. The second shell has room for eight electrons – and the second period has eight elements. The third shell has room for 18 electrons. It fills the first eight spaces, and then the next two electrons are added to the fourth shell. Only then will the ten empty spaces in the third shell begin to fill, and this creates the transition series.

28 **Ni** Nickel	29 **Cu** Copper	30 **Zn** Zinc
46 **Pd** Palladium	47 **Ag** Silver	48 **Cd** Cadmium
78 **Pt** Platinum	79 **Au** Gold	80 **Hg** Mercury
110 **Ds** Darmstadtium	111 **Rg** Roentgenium	112 **Cn** Copernicium

INNER TRANSITION METALS

1

The two lowest rows of the periodic table represent the inner transition metals, although they are

57–71 LANTHANIDES

89–103 ACTINIDES

more commonly known as two series, the lanthanides and actinides. As these elements increase in mass, their additional electrons are positioned not in the outer shell, nor the second one, but in a region of the third layer called the f orbitals.

LANTHANIDES

This series is named after its first member, lanthanum. It contains 15 elements, all metals, which are also frequently described as the rare-earth metals.

57 La LANTHANUM	58 Ce CERIUM	59 Pr PRASEODYMIUM	60 Nd NEODYMIUM	61 Pm PROMETHIUM	62 Sm SAMARIUM
89 Ac ACTINIUM	90 Th THORIUM	91 Pa PROTACTINIUM	92 U URANIUM	93 Np NEPTUNIUM	94 Pu PLUTONIUM

ACTINIDES

This series is named after its first member, actinium. It also contains 15 elements, all radioactive metals. Only two of them, uranium and thorium, are found in any great abundance on Earth.

RARE EARTHS

The lanthanides are called 'rare earths' because they are generally found together in the same obscure ore minerals, such as monazite and ytterbite.

YTTRIUM AND SCANDIUM

Although they do not share the same complex internal electron configuration of the lanthanides, scandium and yttrium, two transition metals, are also regarded as rare earths because they appear in the same ores as the lanthanides.

TECHNOLOGY METALS

Despite their name, rare-earth lanthanides are relatively abundant on Earth. However, they are very difficult to purify. Nevertheless they all have significant uses in high-tech industries, such as optics, electronics and lasers.

WIDE TABLE

The lanthanide and actinide series are more correctly positioned between Group 2 and the transition series. However, to make the table more compact, these 30 elements are more often shown at the bottom of the table.

| 63 Eu EUROPIUM | 64 Gd GADOLINIUM | 65 Tb TERBIUM | 66 Dy DYSPROSIUM | 67 Ho HOLMIUM | 68 Er ERBIUM | 69 Tm THULIUM | 70 Yb YTTERBIUM | 71 Lu LUTETIUM |

| 95 Am AMERICIUM | 96 Cm CURIUM | 97 Bk BERKELIUM | 98 Cf CALIFORNIUM | 99 Es EINSTEINIUM | 100 Fm FERMIUM | 101 Md MENDELEVIUM | 102 No NOBELIUM | 103 Lr LAWRENCIUM |

PRIMORDIAL

All actinides are radioactive and, as a result, most are not found in nature. Any that did exist as primordial elements when Earth's rocks formed have long since decayed out of existence. Only uranium and thorium have half lives long enough for there to be appreciable amounts left. Tiny amounts of primordial plutonium have also been isolated. Actinium, protactinium and neptunium are produced when these other elements decay, and so are found in very small amounts in their ores.

NUCLEAR FUELS

Certain isotopes of thorium, uranium and plutonium are able to undergo nuclear fission. If controlled, this reaction releases heat that is used to drive electricity generators. Other actinides are used in radioisotope thermoelectric generators, where heat from radioactivity is turned directly into electricity.

SYNTHETIC

Usable quantities of most actinides, from uranium up, are synthesized in nuclear reactors, atomic explosions and particle accelerators. Most are then reused to make even larger elements, although a few synthetic elements, such as americium, have found more everyday applications.

THE SHAPES OF ATOMS

1

We generally imagine atoms to be little balls – not solid, necessarily, but spherical blobs of matter. However, quantum mechanics shows that atoms all have a unique shape depending on where their electrons are.

CLOUDY CONCEPT

Electrons are in constant motion around the nucleus. However, it is not possible to locate where an electron is and also know where it is heading to at the same time. Instead, physicists work on the balance of probabilities, and as a result electrons exist in zones that surround the nucleus. The chances are that this is where the electrons are, and so these zones become clouds of charge called orbitals – and the electron is in there somewhere.

F ORBITALS

Elements with more than five shells use f orbitals. The 'f-block' is another name for the lanthanide and actinide series.

D ORBITALS

Elements with more than three electron shells make use of the d orbitals. These never form an atom's outer electron, but exist in the shells beneath. The transition series is called the 'd-block' because their atoms are built from these.

P ORBITALS

The next six electrons fill the lobe-shaped p orbitals. Elements in Groups 3 to 8 form the 'p-block' of the periodic table because their outer electrons are in p orbitals.

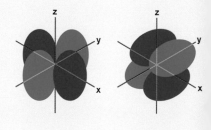

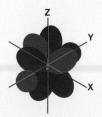

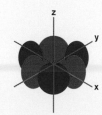

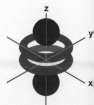

S ORBITALS

The first two electrons in each shell fill the rounded s orbital. Elements in Groups 1 and 2 form the 's-block' of the periodic table because their outer electrons are in the s orbital.

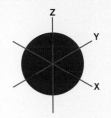

s-block

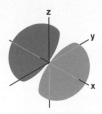
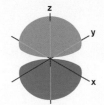
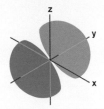

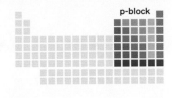

p-block

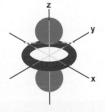
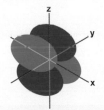
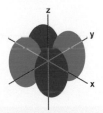

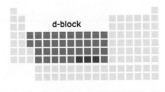

d-block

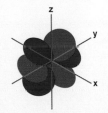
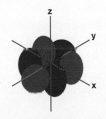

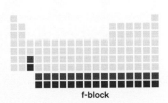

f-block

PHASES OF DISCOVERY

1

There are 118 elements in the current version of the periodic table. Each one has been proven – using different methods – to be a simple substance that cannot be broken down into yet more simple ingredients. The lastest entry, oganesson, was only discovered in 2015, but the process of building the list of elements began at the dawn of history.

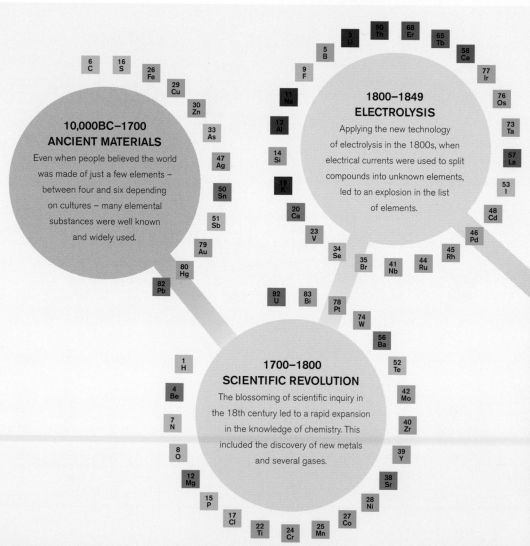

10,000BC–1700
ANCIENT MATERIALS

Even when people believed the world was made of just a few elements – between four and six depending on cultures – many elemental substances were well known and widely used.

1800–1849
ELECTROLYSIS

Applying the new technology of electrolysis in the 1800s, when electrical currents were used to split compounds into unknown elements, led to an explosion in the list of elements.

1700–1800
SCIENTIFIC REVOLUTION

The blossoming of scientific inquiry in the 18th century led to a rapid expansion in the knowledge of chemistry. This included the discovery of new metals and several gases.

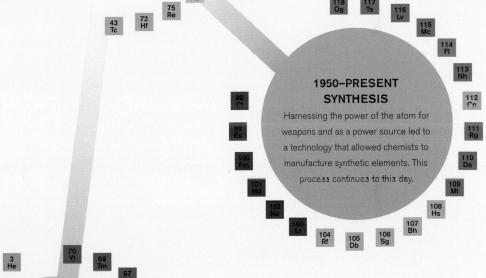

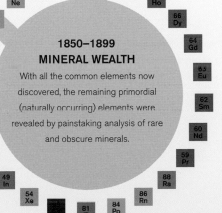

97 Bk 98 Cm 95 Am 94 Pu 93 Np 91 Pa 71 Lu 61 Pm 89 Ac 87 Fr 85 As 75 Re 72 Hf 43 Tc

1900–1949
RADIOACTIVITY

The discovery of the phenomenon of radioactivity at the turn of the 20th century, where atoms of one element can decay into atoms of another, revealed a new set of rare elements.

118 Og 117 Ts 116 Lv 115 Mc 114 Fl 113 Nh 112 Cn 111 Rg 110 Ds 109 Mt 108 Hs 107 Bh 106 Sg 105 Db 104 Rf 103 Lr 102 No 101 Md 100 Fm 99 Es 98 Cf

1950–PRESENT
SYNTHESIS

Harnessing the power of the atom for weapons and as a power source led to a technology that allowed chemists to manufacture synthetic elements. This process continues to this day.

2 He 70 Yt 69 Tm 67 Ho 66 Dy 64 Gd 63 Eu 62 Sm 60 Nd 59 Pr 88 Ra 86 Rn 84 Po 81 Tl 54 Xe 49 In 37 Rb 36 Kr 32 Ge 31 Ga 21 Sc 18 Ar 10 Ne

1850–1899
MINERAL WEALTH

With all the common elements now discovered, the remaining primordial (naturally occurring) elements were revealed by painstaking analysis of rare and obscure minerals.

A HISTORY OF TABLES

1

The periodic table was the brainchild of Dmitri Mendeleev, a Russian chemist. The table layout we use today evolved from a system he developed in 1869, but that was inspired by many earlier attempts to organize the elements.

BEFORE THE FACTS

Neither Mendeleev, nor the chemists who came before, understood anything about atomic structure and how that impacted on the properties of each

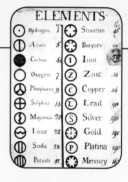

2. AFFINITIES

In 1718, Étienne François Geoffroy listed substances – using their alchemical symbols – according to how they combined or interacted.

4. ATOMS

John Dalton was the first chemist to show that elements worked as atoms. In 1808 he produced this table, which ordered elements by relative weights.

1. ALCHEMY

Alchemists, the wizard-like forebears of modern chemists, organized substances according to their magical properties. They gave them symbols and genders and attributed to them links to planets. This table was put together by Basil Valentine in the 15th century.

3. SIMPLE SUBSTANCES

Antoine Lavoisier drew up a list of *Substances Simple* in 1789. This was one of the first attempts at a table of elements, although it included light, heat and several compounds.

element. They did not know about subatomic particles and had no concept of the atomic number – which is the main factor used to organize elements today. Instead they focused on the relative weights of elements and their chemical properties – most significantly their valence. An element's valence is its combining power, or how many other elements it can link to in compounds.

It was Mendeleev who combined the patterns in atomic weight and valence successfully – and without realizing it, he was ordering elements according to their atomic structure.

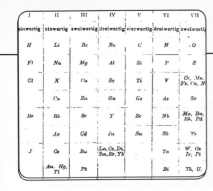

5. TRIADS

In 1817, Johann Döbereiner developed the law of triads, in which he found several sets of three elements that each seemed to share properties. It was published in 1829.

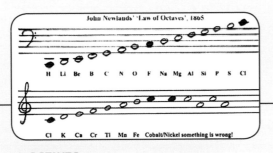

John Newlands' 'Law of Octaves', 1865

6. OCTAVES

In 1864, John Newlands saw that even though every element had a unique relative weight – no two weighed the same – their chemical properties followed a pattern, repeating after every eight elements. He called the groups octaves, and even tried to represent the system using a musical score.

PERIODIC

Finally, in 1869, Mendeleev turned the repeating, or periodic, patterns seen in the valence of elements into the periodic table we know today. However, the first attempt shown here uses columns for the periods – not rows, like we use today.

ALTERNATIVE TABLES

Dmitri Mendeleev came up with his definitive version of the periodic table while playing a game of solitaire, arranging cards in rows and columns – and so the elements in his table are set out in rows and columns, too. But there are other ways to do it.

LAYOUT TRICK

In truth, Mendeleev's version of the periodic table should be much wider than we normally see it. The f-block (which makes up the lanthanide and actinide series) should sit between the s-block (Groups 1 and 2) and the d-block (the transition series). In nearly all cases, that wide section is moved to the bottom to make the table fit into a smaller space and also make it easier to read. However, a circular table can get around this problem in another way.

MULTIPLE SPIRAL

In 1964, Theodor Benfey designed a multi-centred spiral version of the periodic table. The 'periodic divide' shows where a new period (or row in Mendeleev's version) begins. The s-block and p-block are arranged in a spiral around hydrogen.

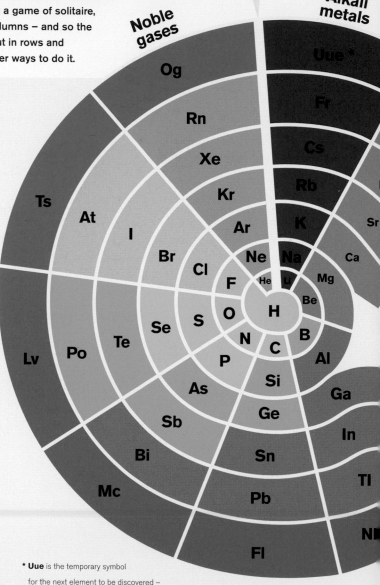

Periodic divide

Noble gases

Alkali metals

* **Uue** is the temporary symbol for the next element to be discovered – Ununennium 119. It hasn't been made yet.

FUTUREPROOF

Benfey's version also has room for a new 'g-block' called the superactinide series. These are super-massive elements that have yet to be made in the laboratory.

Superactinides

Lathanides & actinides

Transition metals

PENINSULAE

The d-block (transition metals) forms a 'peninsula' off to the side, and another 'peninsula' shows the f-block (lanthanides and actinides).

2

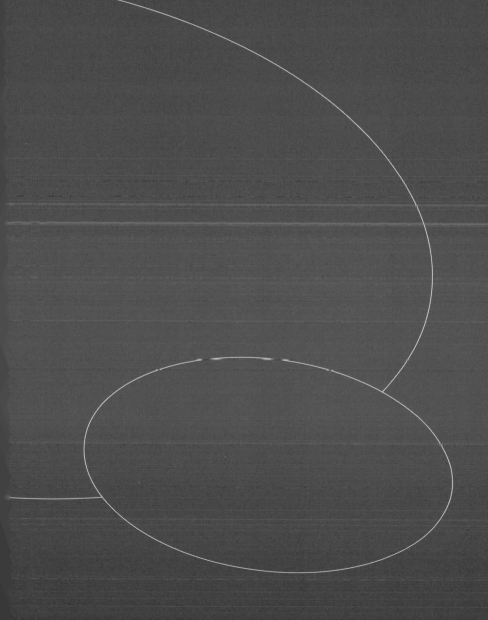

THE BIG PICTURE

COUNTING ATOMS

Atoms are too small to see even with the most powerful microscope, so that makes them difficult to count one by one. Instead, chemists group them together in a unit called a mole. The number of atoms in one mole is 602,214,179,000,000,000,000,000.

One mole of seconds is one million times longer than the age of the Universe.

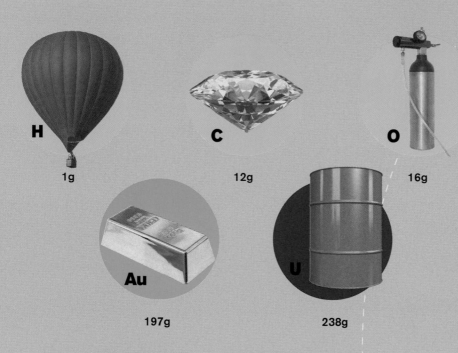

H
1g

C
12g

O
16g

Au
197g

U
238g

MOLES AND ATOMS

We can count the atoms in these elements because the atoms of every element have a unique mass. This is based on the number of protons and neutrons it has (the electrons are too small to count). Hydrogen atoms have one proton, while carbon atoms have six protons and six neutrons. Therefore, the relative atomic mass (RAM) of hydrogen is 1, and the RAM of carbon is 12. In other words, a carbon atom is always 12 times heavier than a hydrogen atom. Chemists have decided, therefore, that a mole of an element's atoms is simply its RAM measured in grams.

602,214,179,000

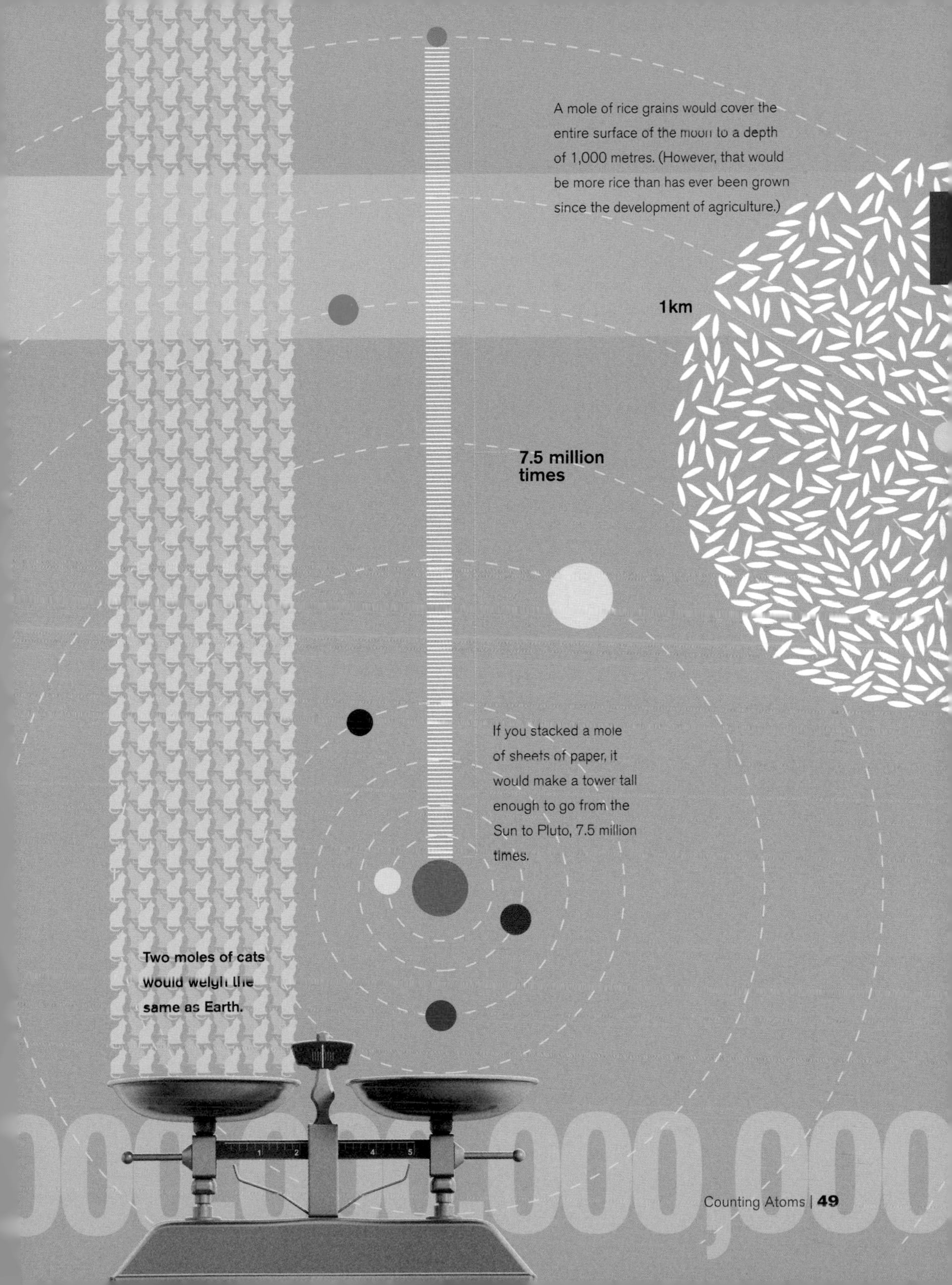

A mole of rice grains would cover the entire surface of the moon to a depth of 1,000 metres. (However, that would be more rice than has ever been grown since the development of agriculture.)

1km

7.5 million times

If you stacked a mole of sheets of paper, it would make a tower tall enough to go from the Sun to Pluto, 7.5 million times.

Two moles of cats would weigh the same as Earth.

THE SIZE OF AN ATOM

Atoms get heavier as their atomic number increases. This is because the number of particles inside them is going up. However, the size of the atom – measured as the radius, or the distance from the nucleus to the outer electron shell – does not increase in the same way.

SIZE TREND

The radius of an atom shrinks as we move across each period. As the number of protons goes up, their increasing positive charge pulls harder on the outer layer of electrons. That draws the electrons closer to the nucleus. At the start of each period, a new layer of electrons is formed, making the atom balloon out again.

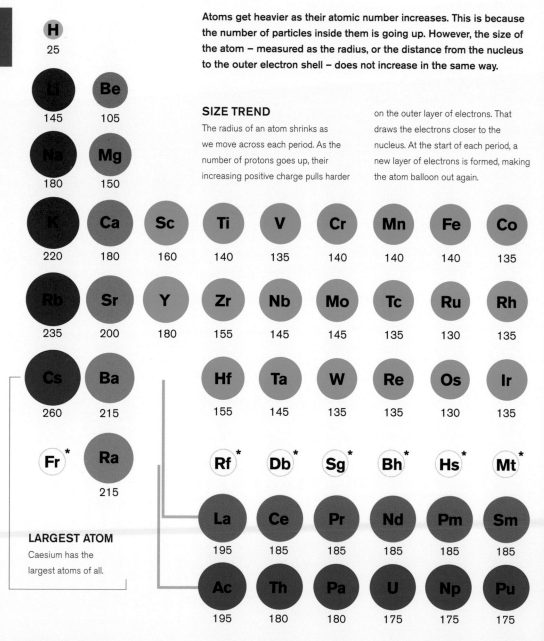

H 25								
Li 145	**Be** 105							
Na 180	**Mg** 150							
K 220	**Ca** 180	**Sc** 160	**Ti** 140	**V** 135	**Cr** 140	**Mn** 140	**Fe** 140	**Co** 135
Rb 235	**Sr** 200	**Y** 180	**Zr** 155	**Nb** 145	**Mo** 145	**Tc** 135	**Ru** 130	**Rh** 135
Cs 260	**Ba** 215		**Hf** 155	**Ta** 145	**W** 135	**Re** 135	**Os** 130	**Ir** 135
Fr *	**Ra** 215		**Rf** *	**Db** *	**Sg** *	**Bh** *	**Hs** *	**Mt** *
			La 195	**Ce** 185	**Pr** 185	**Nd** 185	**Pm** 185	**Sm** 185
			Ac 195	**Th** 180	**Pa** 180	**U** 175	**Np** 175	**Pu** 175

LARGEST ATOM

Caesium has the largest atoms of all.

RADIUS MEASUREMENT

An atom's size actually frequently varies as it takes in and gives out energy. As a result, the radius value is calculated by measuring the distance between the nuclei of two atoms bonded together and then halving that figure. This gives a stable, verifiable value.

NO NOBLES

The noble gases do not form bonds, and so the radii of their atoms cannot be measured in the same way.

He *

UNITS

The radius of an atom is measured in picometres (pm). One picometre is a trillionth of a metre.

B	C	N	O	F	Ne *
85	70	65	60	50	

Al	Si	P	S	Cl	Ar *
125	110	100	100	100	

Ni	Cu	Zn	Ga	Ge	As	Se	Br	Kr *
135	135	135	130	125	115	115	115	

Pd	Ag	Cd	In	Sn	Sb	Te	I	Xe *
140	160	155	155	145	145	140	140	

Pt	Au	Hg	Tl	Pb	Bi	Po	At *	Rn *
135	135	150	190	180	160	190		

Ds *	Rg *	Cn *	Nh *	Fl *	Mc *	Lv *	Ts *	Og *

Eu	Gd	Tb	Dy	Ho	Er	Tm	Yb	Lu
185	180	175	175	175	175	175	175	175

Am	Cm *	Bk *	Cf *	Es *	Fm *	Md *	No *	Lr *
175								

* NO DATA: The radii of rare and short-lived radioactive elements have not been measured.

DENSITY TRENDS

Density is the measure of how much mass is held within a particular volume. It seems sensible to think that the elements with the biggest, heaviest atoms would always be the most dense. However, that would be a mistake.

TABLE TALK

The least-dense element in nature is hydrogen and the densest is iridium (with osmium a very close second). The artificial elements, such as meitnerium, are thought to be even denser, but very little of these substances have been produced to date. Density follows a trend through the table. It increases down the groups, and rises in the middle section, especially the transition series. However, the elements at the start and end of the periods have a lower density.

LOG SCALE

This graphic represents the density of each element by the width of a circle according to a logarithmic scale. A doubling of the size of a circle indicates a ten-fold increase in density.

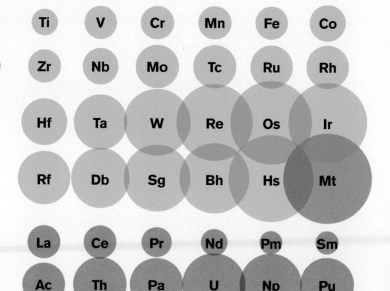

PUSH AND PACK

The density of an element does not simply depend on how heavy its atoms are. It is also important to consider how large the atom is and, most importantly, how closely the atoms can pack together – or if they can bond to one another at all. The least dense elements are the gases. The atoms in these elements do not cluster together and so will spread out to occupy a greater volume. In the solid (and liquid) elements, when atoms get close to one another, their electrons push away from each other; a negative charge always repels another negative charge,

As this table shows, some atoms are more pushy than others. Elements on the left of the table have a weak negative charge on their surfaces, but have very large atoms. Even when packed closely, these atoms do not weigh that much. The elements on the right, meanwhile, have compact atoms, but that is combined with a powerful

negative charge on the outside. As a result they push each other away and fill a larger volume. Central elements have more compact and heavy atoms, but they also lack powerful negative charges at their surface. Therefore, these elements pack their atoms most densely of all.

He
Ne
B   C   N   O   F
Al   Si   P   S   Cl   Ar
Ni   Cu   Zn   Ga   Ge   As   Se   Br   Kr
Pd   Ag   Cd   In   Sn   Sb   Te   I   Xe
Pt   Au   Hg   Tl   Pb   Bi   Po   At   Rn
Ds   Rg   Cn   Nh   Fl   Mc   Lv   Ts   Og
Eu   Gd   Tb   Dy   Ho   Er   Tm   Yb   Lu
Am   Cm   Bk   Cf   Es   Fm   Md   No   Lr

DENSITY COMPARISON

2

Density is calculated by dividing an object's mass (or weight) by its volume. We have a simple benchmark for density: water. The easiest way to comprehend an element's density is to compare it with that of water. If it floats, it is less dense – and if it sinks, it is more dense.

DENSITY BY ATOMIC NUMBER

This chart shows how density changes with atomic number (and atomic mass). Each hump in the line indicates a period, or row, of the periodic table. The densities of elements that share a period rise as they approach the middle, and then drop back down, reaching their lowest point, or nadir, at the end of each period. The nadirs are filled by the Group 8 noble gases. One exception is hydrogen, which is the first and least-dense element.

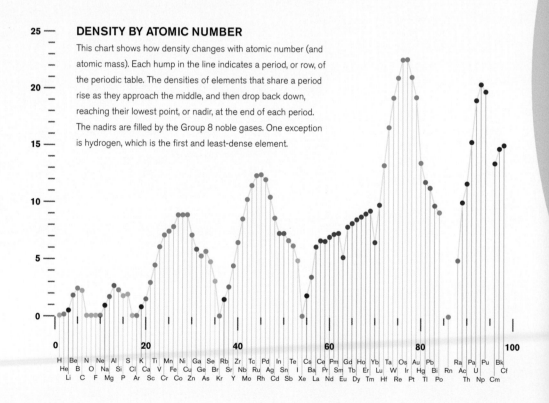

WATER AND DENSITY

Mass is measured in kilograms, and volume is easily measured in litres. A litre is 1,000 millilitres, and one millilitre is the same as one cubic centimetre. The kilogram is cleverly designed so that one litre of water weighs one kilogram. As a result, the density of water is one kilogram per litre (and one gram per cubic centimetre). All densities are calculated in the same way. In this chart, we compare one millilitre of water with the same weight of a range of elements.

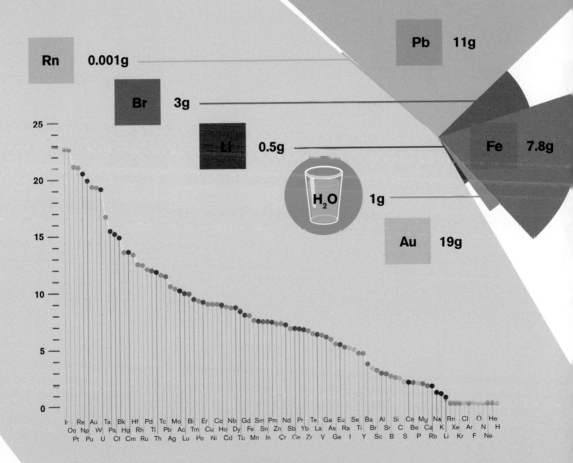

Rn	0.001g
Br	3g
Li	0.5g
Pb	11g
Fe	7.8g
H₂O	1g
Au	19g

DENSITY BY VALUE

This chart shows the elements arranged by density. The d-block and f-block metals dominate the left side, and the gases are to the right.

ELEMENTS ON EARTH

Astronomers have discovered that the elements found on Earth are seen all over the Universe. However, they are not evenly spread, and that includes on Earth itself. There is a marked difference between the elemental composition of different regions of our planet.

Earth is divided into three distinct layers: the core, the mantle and the crust. The core is dominated by heavier metallic elements, which sank to the centre while the young Earth was a seething blob of molten rock. The centre of the Earth is still molten. At the same time, lighter elements, such as silicon, aluminium and oxygen, drifted to the surface area. As the outer region of Earth cooled, these elements solidified and formed the basis for the solid, rocky crust. The surface of Earth is also covered by oceans of water and then a blanket of air, both with their own distinctive elemental compositions.

O   46%

Si   27%

N_2 78%

Ar   0.9%

Al   8.2%

Fe   6.3%

O_2 20.9%

O   85.7%

H   10.8%

Mg   0.1%

Ca   5.0%

Cl   1.9%

Mg   2.9%

Na   1.1%

Na   2.3%

K   1.5%

Atmosphere

Sea

Element proportions by mass

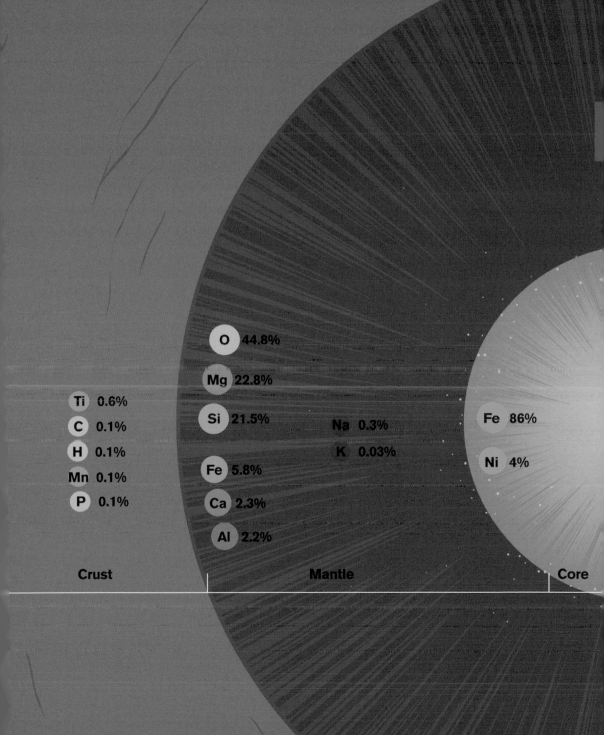

Ti 0.6%
C 0.1%
H 0.1%
Mn 0.1%
P 0.1%

O 44.8%
Mg 22.8%
Si 21.5%
Fe 5.8%
Ca 2.3%
Al 2.2%

Na 0.3%
K 0.03%

Fe 86%
Ni 4%

Crust

Mantle

Core

HUMAN BODY COMPOSITION

2

Your body is a made from chemicals just like everything else. Until the 1820s, it was assumed that some non-chemical 'vital force' drove the processes of a living body. Then it was found that, although complex, the body used the same kinds of chemical reactions as seen elsewhere.

SQUARED AWAY

Each one of these squares represents one per cent of the body by weight.

Ninety-four per cent of the body's weight is made up of oxygen, carbon and hydrogen. These are the elements used in sugars, starches and fats. They are also the dominant components in proteins.

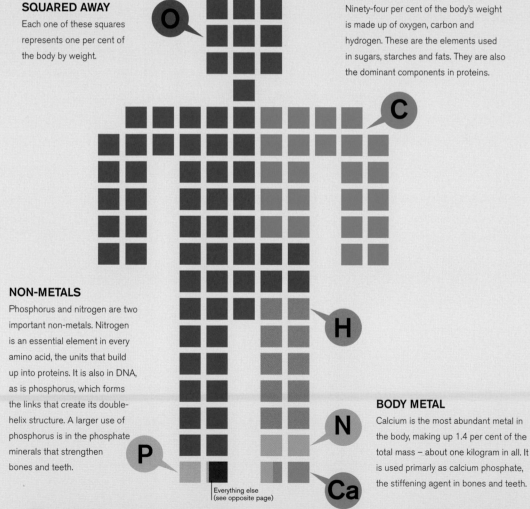

NON-METALS

Phosphorus and nitrogen are two important non-metals. Nitrogen is an essential element in every amino acid, the units that build up into proteins. It is also in DNA, as is phosphorus, which forms the links that create its double-helix structure. A larger use of phosphorus is in the phosphate minerals that strengthen bones and teeth.

Everything else
(see opposite page)

BODY METAL

Calcium is the most abundant metal in the body, making up 1.4 per cent of the total mass – about one kilogram in all. It is used primarily as calcium phosphate, the stiffening agent in bones and teeth.

BODY OF ELEMENTS

The human body contains 60 elements. More than 99 per cent of the human body's weight is made up of just six elements: oxygen, carbon, hydrogen, nitrogen, phosphorus and calcium. The next 0.85 per cent is composed of potassium, sulphur, sodium, chlorine and magnesium, which means the other 49 elements – known as trace elements – make up just 0.15 per cent, or ten grams.

= **Important function in body (see left)**

= **Known function in body**

= **No function in body**

= **Possible function in body**

= **Not present in body**

WITH A PURPOSE

Eighteen of the trace elements have a known or suspected function in the body. They include arsenic, cobalt and even fluorine, which are all deadly in large amounts.

ALONG FOR THE RIDE

Thirty-one of the trace elements have no known function and are only present in minute quantities. They include gold, caesium and uranium, and are probably present simply as impurities that arrive consistently via our food.

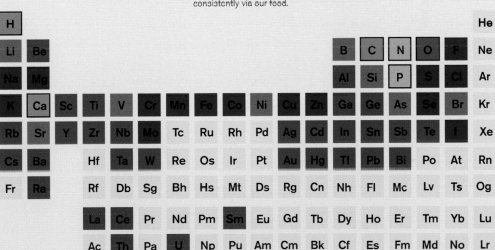

CHANGING STATES

Every element has a standard state, either solid, liquid or gas, in standard conditions (25°C at 1 atmosphere of pressure). However, that can change by adding or removing heat energy to make the element melt, freeze, boil or condense.

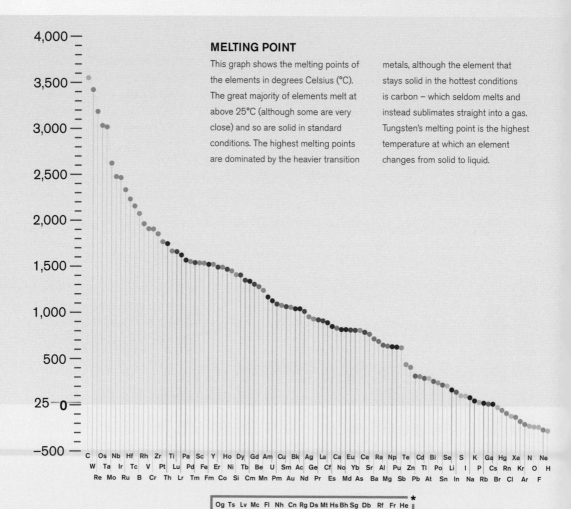

MELTING POINT

This graph shows the melting points of the elements in degrees Celsius (°C). The great majority of elements melt at above 25°C (although some are very close) and so are solid in standard conditions. The highest melting points are dominated by the heavier transition metals, although the element that stays solid in the hottest conditions is carbon – which seldom melts and instead sublimates straight into a gas. Tungsten's melting point is the highest temperature at which an element changes from solid to liquid.

* No data available

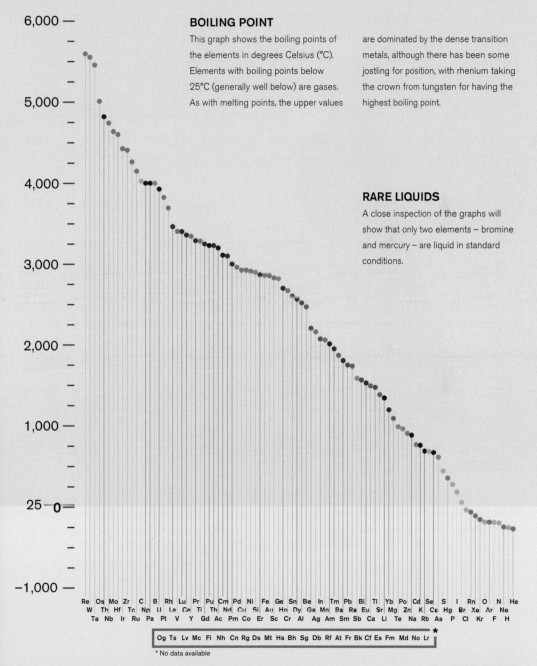

BOILING POINT

This graph shows the boiling points of the elements in degrees Celsius (°C). Elements with boiling points below 25°C (generally well below) are gases. As with melting points, the upper values are dominated by the dense transition metals, although there has been some jostling for position, with rhenium taking the crown from tungsten for having the highest boiling point.

RARE LIQUIDS

A close inspection of the graphs will show that only two elements – bromine and mercury – are liquid in standard conditions.

6,000 —

5,000 —

4,000 —

3,000 —

2,000 —

1,000 —

25 — 0

−1,000 —

Re Os Mo Zr C B Rh Lu Pr Pu Cm Pd Ni Fe Ge Sn Be In Tm Pb Bi Tl Yb Po Cd Se S I Rn O N He
W Th Hf Tc Np Ll La Ce Ti Th Nd Cu Si Au Ho Dy Ga Mn Ba Eu Sr Mg Zn K Ca Hg Br Xe Ar Ne
Ta Nb Ir Ru Pa Pt V Y Gd Ac Pm Co Er Sc Cr Al Ag Am Sm Sb Ca Li Te Na Rb As P Cl Kr F H

Og Ts Lv Mc Fl Nh Cn Rg Ds Mt Hs Bh Sg Db Rf At Fr Bk Cf Es Fm Md No Lr *

* No data available

REACTIVITY

Chemical reactions are driven by an element attempting to fill (or empty) its outer electron shell. Non-metals do this by adding electrons taken from other atoms, while metals give away their outer electrons. The most reactive elements are the ones that give away electrons and add new ones most readily.

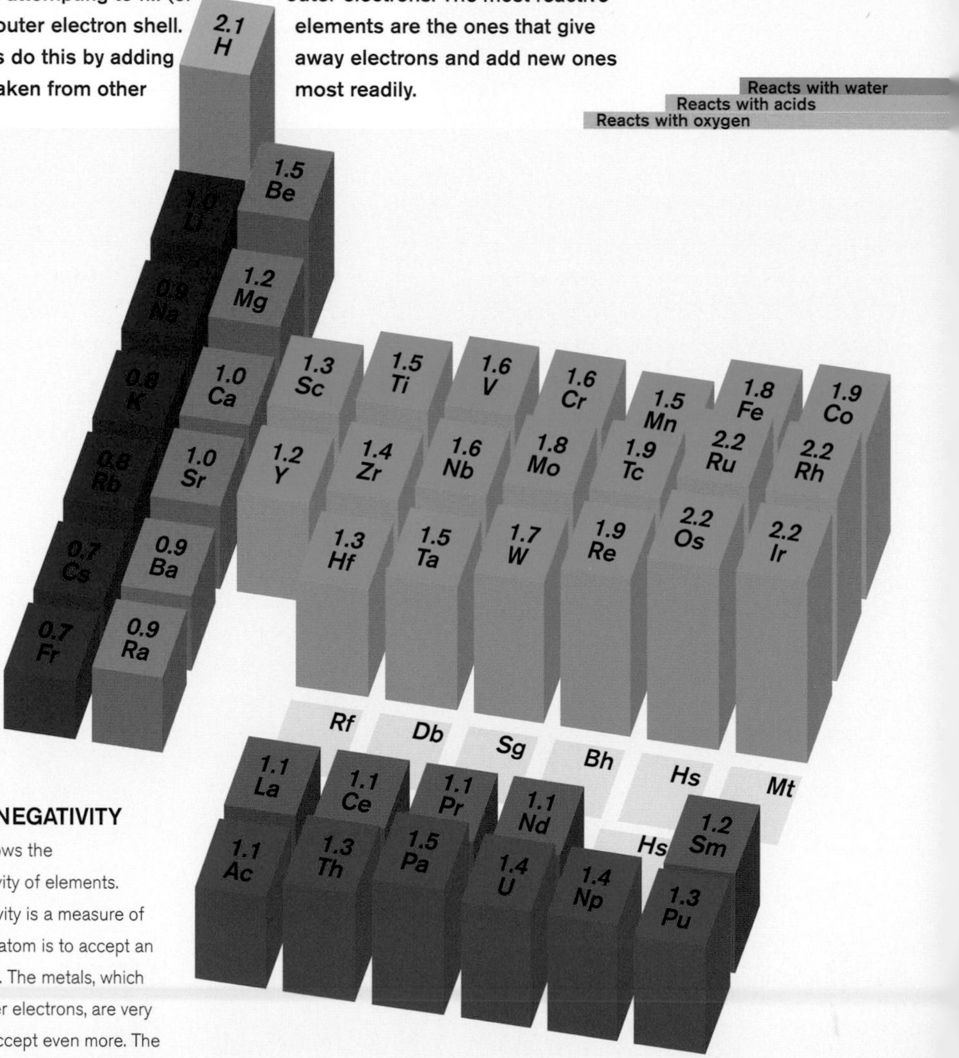

Reacts with water
Reacts with acids
Reacts with oxygen

2.1
H

1.5
Be

1.0
Li

0.9
Na

1.2
Mg

0.8
K

1.0
Ca

1.3
Sc

1.5
Ti

1.6
V

1.6
Cr

1.5
Mn

1.8
Fe

1.9
Co

0.8
Rb

1.0
Sr

1.2
Y

1.4
Zr

1.6
Nb

1.8
Mo

1.9
Tc

2.2
Ru

2.2
Rh

0.7
Cs

0.9
Ba

1.3
Hf

1.5
Ta

1.7
W

1.9
Re

2.2
Os

2.2
Ir

0.7
Fr

0.9
Ra

Rf   Db   Sg   Bh   Hs   Mt

1.1
La

1.1
Ce

1.1
Pr

1.1
Nd

Hs

1.2
Sm

1.1
Ac

1.3
Th

1.5
Pa

1.4
U

1.4
Np

1.3
Pu

ELECTRONEGATIVITY

This chart shows the electronegativity of elements. Electronegativity is a measure of how likely an atom is to accept an extra electron. The metals, which have few outer electrons, are very reluctant to accept even more. The non-metals are nearer to a full set of outer electrons and so are more eager to collect another.

REACTIVITY SERIES

This chart shows the relative reactivity of certain common metals. The most reactive will react with cold water, acids and oxygen. The least reactive do none of these things.

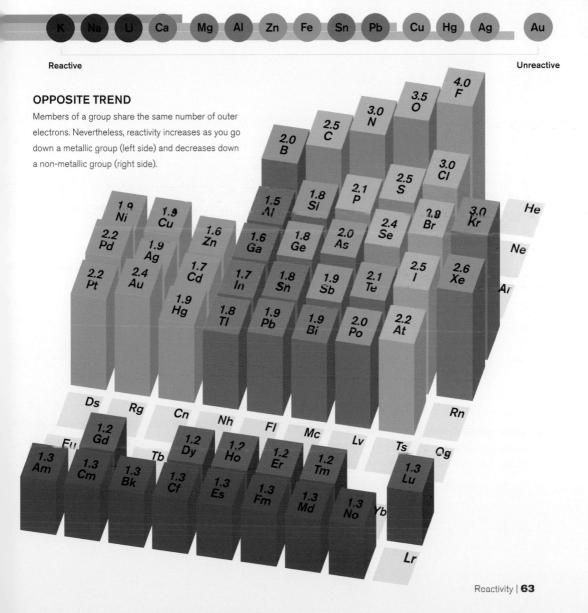

K Na Li Ca Mg Al Zn Fe Sn Pb Cu Hg Ag Au

Reactive   Unreactive

OPPOSITE TREND

Members of a group share the same number of outer electrons. Nevertheless, reactivity increases as you go down a metallic group (left side) and decreases down a non-metallic group (right side).

HARDNESS

It is difficult to quantify the hardness of a substance and there are several competing systems that aim to do so. The simplest is Mohs' Scale, which compares the hardness of a material against ten index minerals, all found pure in nature.

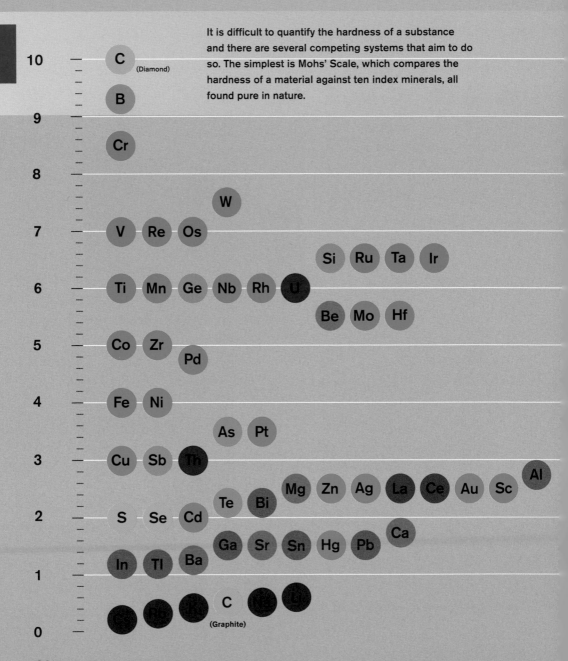

SCRATCH TEST

Mohs' Scale works by scratching the index mineral against a substance, in this case pure solid elements. You then look for a mark. If the index mineral gets scratched, then the element is the harder substance and you move to the next comparator in the scale. Eventually the pure element will get a scratch and you have a rough hardness value. A range of other minerals can then be used to get a more exact figure.

Diamond

Corundum

Topaz

Quartz

Feldspar

Apatite

Fluorite

Calcite

Gypsum

Talc

COMPARE NOT MEASURE

Mohs' Scale is easy and effective but does not give true values of relative hardness. Talc (hardness 1) is not ten times softer than diamond (hardness 10), but actually thousands of times so. Only solid elements can be tested for hardness, but many solid elements are too rare or radioactive to be tested this way.

Cm	Am	Pu	Np	Pa	Ac	Rn	At	Po	Xe	I	Tc	Kr	Br	Ar	Cl	P	F	O	N	He	H
Cf	Bk	Og	Ts	Lv	Mc	Fl	Nh	Cn	Rg	Ds	Mt	Hs	Bh	Sg	Db	Rf	Lr	No	Md	Fm	Es

*

* No data available

STRENGTH

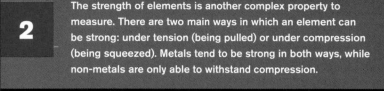

2

The strength of elements is another complex property to measure. There are two main ways in which an element can be strong: under tension (being pulled) or under compression (being squeezed). Metals tend to be strong in both ways, while non-metals are only able to withstand compression.

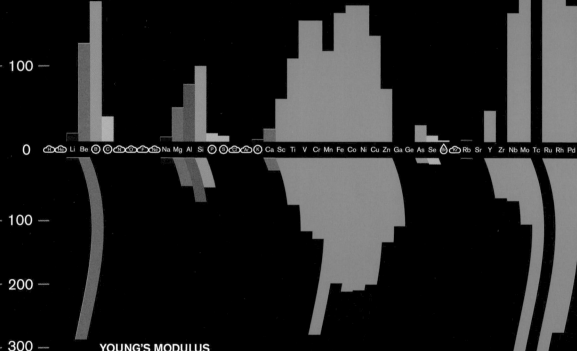

— 200 —

— 100 —

0

— 100 —

— 200 —

— 300 —

YOUNG'S MODULUS

This is a measure of how much a material stretches before it breaks. There are two stages of stretching: elastic, where the material deforms temporarily but returns to its original shape when the force is removed; and plastic, where the deformation is permanent. The yield point marks the transition from elastic to plastic deformation. No elements are truly elastic in the common meaning of the word. Ductile metals, such as gold, will stretch and not break until well beyond their yield points, while more brittle ones such as iron will barely stretch but break soon after they begin to deform.

— 400 —

— 500 —

BULK MODULUS

This is a measure of compressive strength that compares how much a solid decreases in volume as the pressure squeezing its surface increases. Specifically, the modulus relates how much pressure is needed to reduce the volume by one per cent.

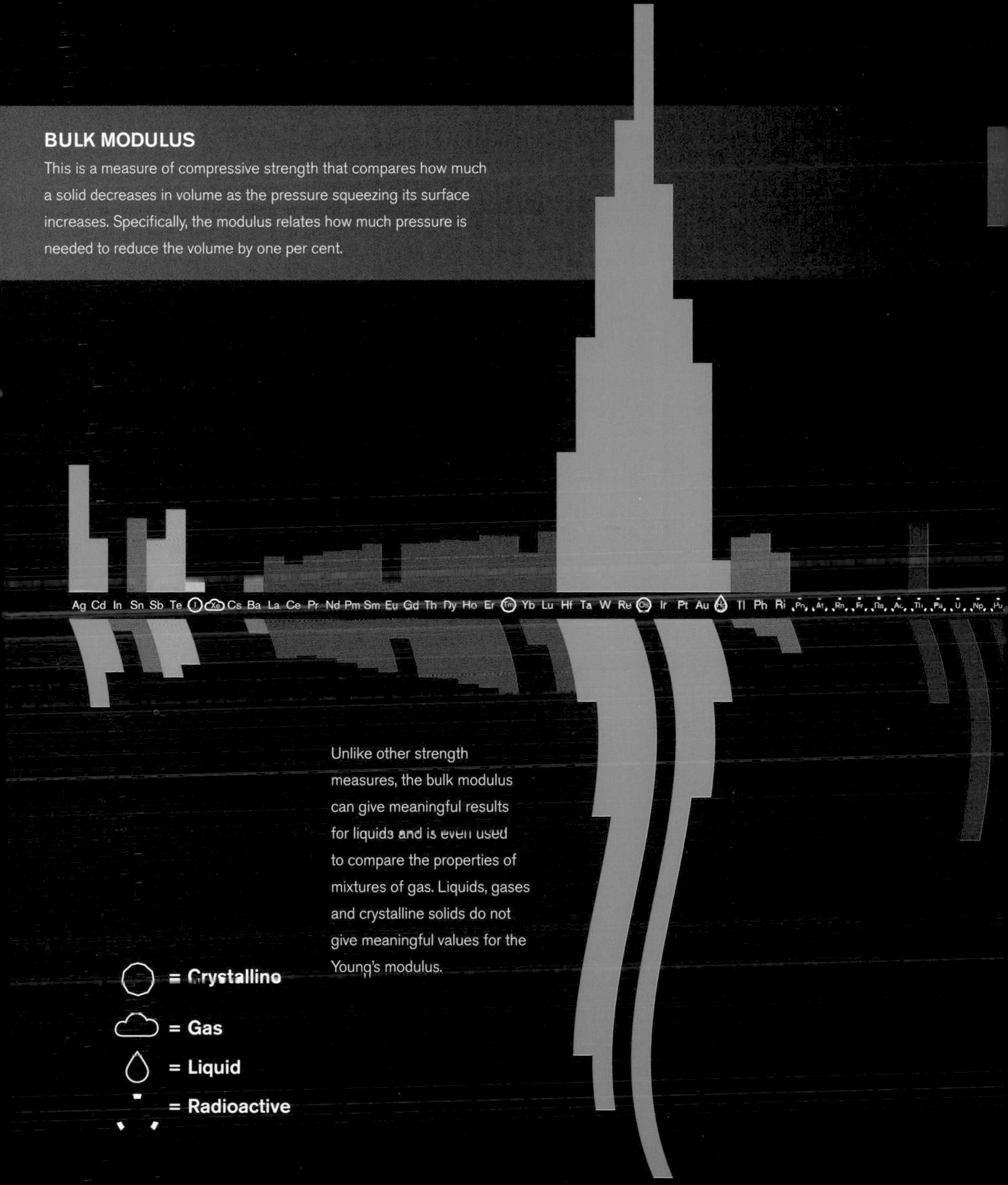

Ag Cd In Sn Sb Te I Xe Cs Ba La Ce Pr Nd Pm Sm Eu Gd Th Dy Ho Er Tm Yb Lu Hf Ta W Re Os Ir Pt Au Hg Tl Pb Bi Po At Rn Fr Ra Ac Th Pa U Np Pu

Unlike other strength measures, the bulk modulus can give meaningful results for liquids and is even used to compare the properties of mixtures of gas. Liquids, gases and crystalline solids do not give meaningful values for the Young's modulus.

= Crystalline

= Gas

= Liquid

= Radioactive

CONDUCTIVITY

2

There are two forms of conductivity: electrical and thermal. If an element is a good conductor of one, it is likely to be good at conducting the other. However, there are some exceptions to that rule, exceptions that have had a big impact on science, technology and everyday life.

HANDLING HEAT

Heat is really the motion of atoms. When an element gets hotter, its atoms move with more energy. In the case of a solid element, where the atoms are bonded together, that means they are vibrating back and forth more violently than before. A good conductor of heat, therefore, is a substance in which that oscillatory motion can be transferred from one atom to its neighbours.

CARRYING CURRENT

Electricity is a flow of charge through a material. This is generally carried as a wave of negatively charged electrons that surges through a substance. Metals are the best conductors of electricity because their atoms have few outer electrons, which readily free themselves from the atom and move as a current. Non-metal atoms hold their electrons much more tightly, and so need a larger electrical shove, or voltage, to get them to conduct electrical currents.

47 Ag SILVER	29 Cu COPPER

Silver and copper are the best conductors of all. Metals tend to be the best conductors of heat because their atoms are more free to move around and transfer energy from one to the other. Silver and copper atoms only have one outer electron, which is easily freed up to carry electricity.

54 Xe XENON	86 Rn RADON

The dense gases xenon and radon are the worst conductors of heat, because their heavy, sluggish atoms are not connected at all. All gases are non-metals and as such are extremely poor conductors of electricity.

14 Si SILICON	32 Ge GERMANIUM

Silicon and germanium stand out from the other elements as being relatively good heat conductors like metals, but relatively poor electrical conductors like non-metals. With four outer electrons, they sit halfway between typical metals and typical non-metals and are thus known as semimetals. Both elements are the main sources of semiconducting materials, which can be switched from being electrical insulators to conductors. This switching forms the basis of electronics and computer technology. Semiconductors typically contain tiny amounts of dopants, such as tin, boron and arsenic, which enhance the ability to carry current.

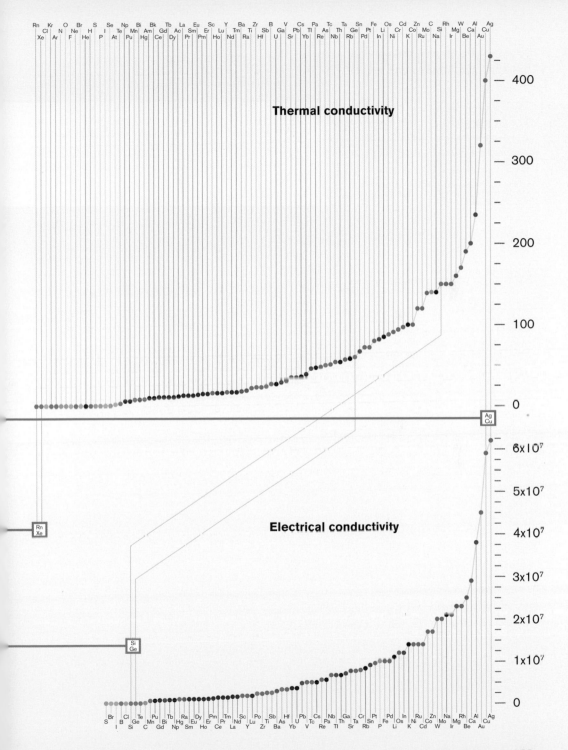

Thermal conductivity

Electrical conductivity

MAGNETISM

All elements exhibit a form of magnetism – it is just too weak to notice or take account of in everyday life. There are four kinds of magnetic effects: ferromagnetism, paramagnetism, diamagnetism and antiferromagnetism.

ACCUMULATED ACTION

All atoms of all elements produce magnetic force fields. However, these tiny fields are generally orientated in random directions, and so the cumulative force produced is cancelled out into nearly nothing. However, if the element is placed in a magnetic field, its atoms will switch orientation to line up with it. This reveals the element's magnetic character.

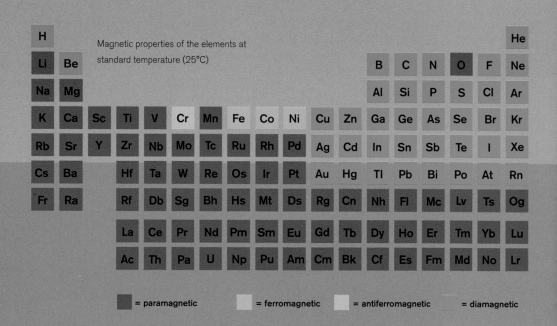

Magnetic properties of the elements at standard temperature (25°C)

■ = paramagnetic   ■ = ferromagnetic   ■ = antiferromagnetic   ■ = diamagnetic

Paramagnetic: atoms align with the field and are attracted to its source. When the source is removed, the atoms re-jumble and lose their magnetic pull.

Ferromagnetic: atoms align with the field and are attracted to its source. The atoms remain aligned when the field is removed and continue to exhibit magnetic properties permanently.

Antiferromagnetic: atoms align with the field, although half are orientated towards it and and half away. The combined forces from the atoms cancels out to nothing.

Diamagnetic: atoms align with the field but repel the source. All magnetic effects are lost once the field is removed.

TEMPERATURE DEPENDENT

An element's magnetic characteristics are temperature dependant. For example, holmium atoms produce the strongest magnetic pull of any element, but only become ferromagnetic at −254°C.

Paramagnetic material

Ferromagnetic material

Antiferromagnetic material

Diamagnetic material

Normal state

Magnetic field present

Magnetic field removed

SPECTRA

Light is radiation produced by atoms. An atom gives out light when its electrons lose energy, and every element produces light at a very specific set of wavelengths, or colours. These unique atomic spectra can be used to identify atoms simply by the light they emit.

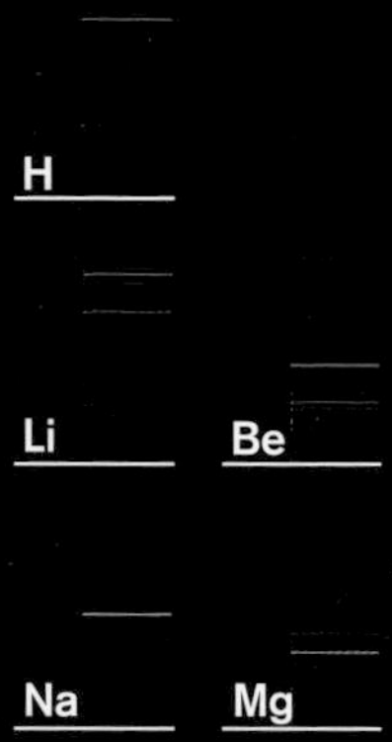

FLAME TEST

The simplest way to identify an element is to set it on fire. The light from its flame is made up of the telltale colours of its spectra.

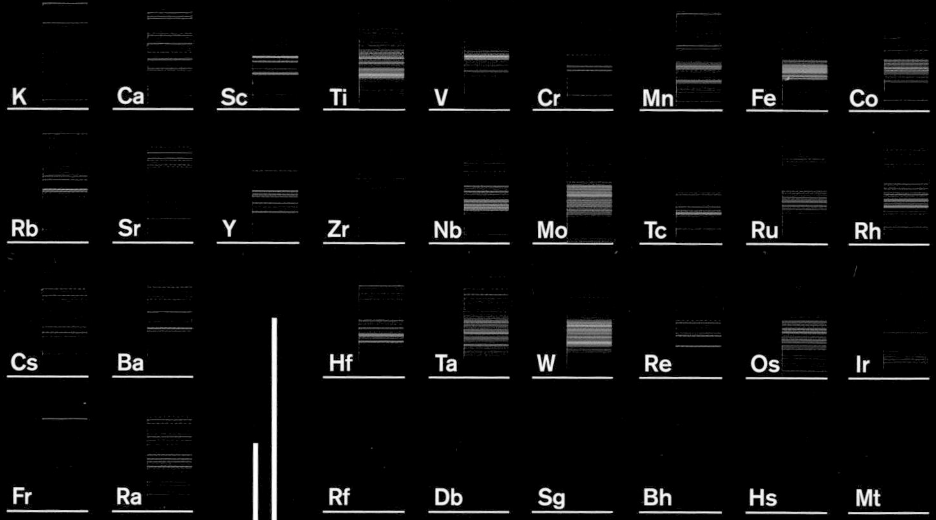

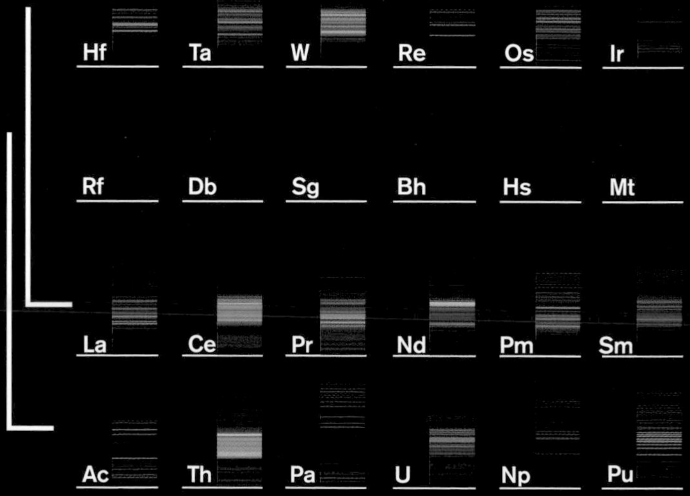

ABSORPTION

Atoms give out light when they are hot, but will absorb it when they are cold. Astronomers can identify the elements in clouds of gas out in space from the colours they absorb as starlight shines through them.

H

Li   Be

Na   Mg

K   Ca   Sc   Ti   V   Cr   Mn   Fe   Co

Rb   Sr   Y   Zr   Nb   Mo   Tc   Ru   Rh

Cs   Ba   Hf   Ta   W   Re   Os   Ir

Fr   Ra   Rf   Db   Sg   Bh   Hs   Mt

La   Ce   Pr   Nd   Pm   Sm

Ac   Th   Pa   U   Np   Pu

DISCOVERY TOOL

Many elements, most notably helium, were first dicovered by the light of their spectra. Several, such as rubidium, caesium and thallium, were named after the colours they produce (red, blue and green respectively).

He

B   C   N   O   F   Ne

Al   Si   P   S   Cl   Ar

Ni   Cu   Zn   Ga   Ge   As   Se   Br   Kr

Pd   Ag   Cd   In   Sn   Sb   Te   I   Xe

Pt   Au   Hg   Tl   Pb   Bi   Po   At   Rn

Ds   Rg   Cn   Nh   Fl   Mc   Lv   Ts   Og

Eu   Gd   Tb   Dy   Ho   Er   Tm   Yb   Lu

Am   Cm   Bk   Cf   Es   Fm   Md   No   Lr

ORIGINS OF ELEMENTS

2

The elements have not existed forever. They were made in nuclear reactions that forced smaller atoms to fuse together into larger ones.

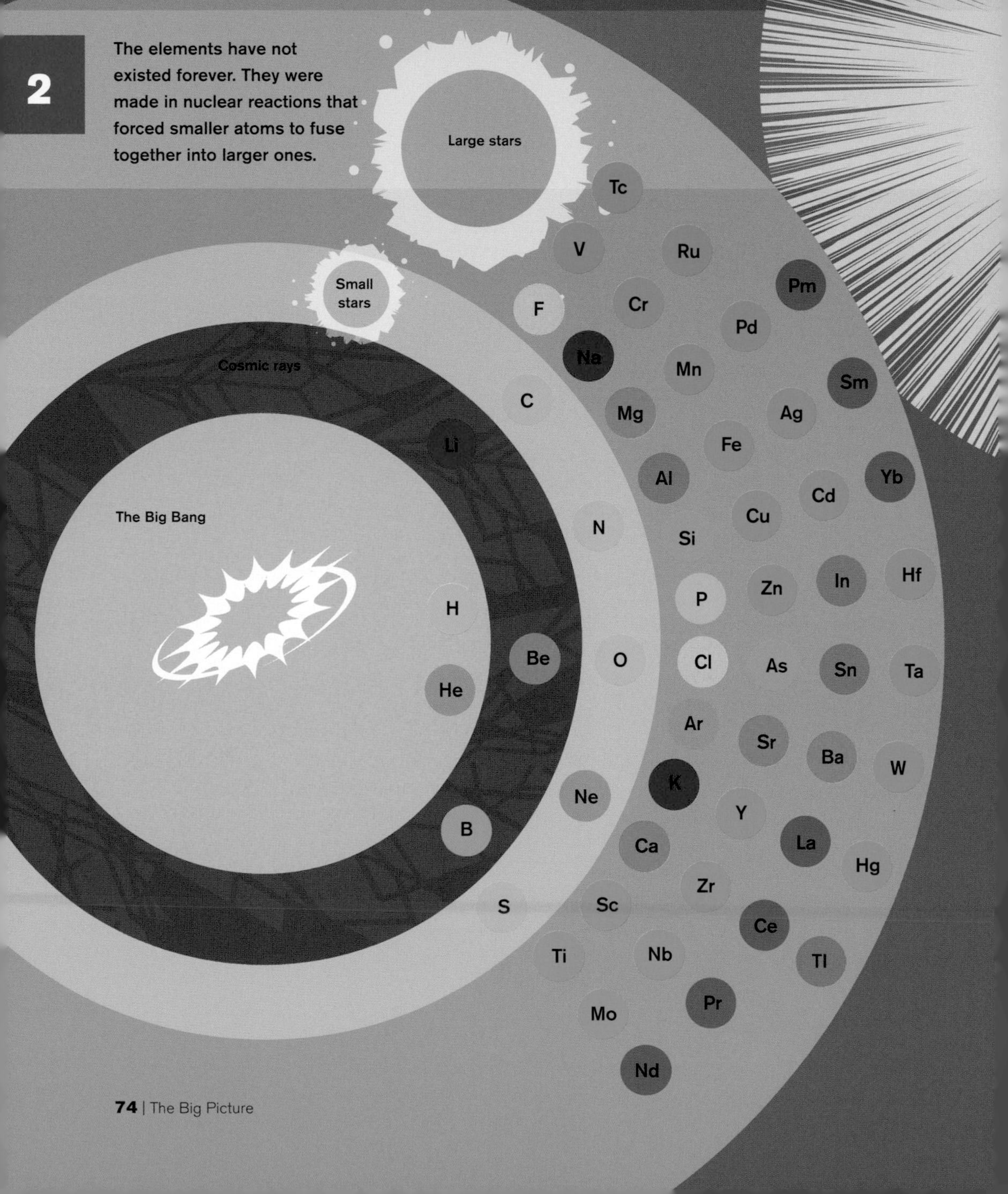

Large stars

Small stars

Cosmic rays

The Big Bang

Tc

V   Ru

F   Cr   Pm

Na   Pd

Mn   Sm

C   Mg   Ag

Li   Fe

Al   Yb

N   Cu   Cd

Si

P   Zn   In   Hf

H   Be   O   Cl   As   Sn   Ta

He   Ar

Sr   Ba   W

K   Y

B   Ne   La   Hg

Ca

Zr   Ce

S   Sc   Tl

Ti   Nb

Pr

Mo

Nd

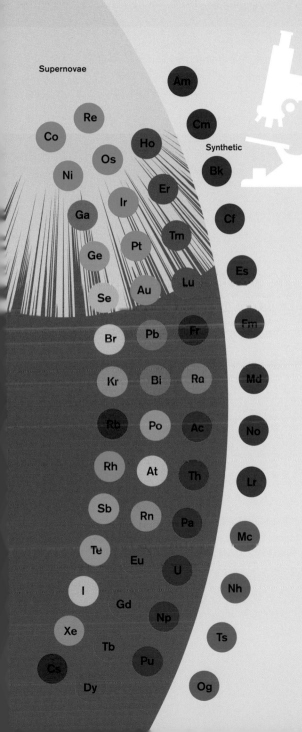

Supernovae

Synthetic

IN THE BEGINNING

Nuclei of hydrogen, the simplest atom formed after the Big Bang, began fusing into helium.

Once the Universe expanded, cosmic rays of high-speed hydrogen and helium nuclei fused into larger elements, such as lithium, beryllium and boron.

ATOM FACTORIES

Stars begin as vast balls of hydrogen that are heated and compressed by their own gravity. Nuclear fusion at the core produces helium, and when the hydrogen runs out, the helium begins to fuse into heavier elements. All elements heavier than boron have been formed inside a star.

A star like our Sun can produce elements as large as neon. Then it will expand into a giant star capable of making heavier elements. The largest stars of all die in immense explosions called supernovae. These enormously violent events make the heaviest naturally occurring elements.

IN THE LAB

Elements larger than plutonium may be made in supernovae, but they do not last long and are not seen anywhere in nature. Instead they are synthesized using nuclear reactors and particle accelerators.

UNIVERSAL ABUNDANCE

Some elements are more common in the Universe than others. The general trend is that the elements with low atomic numbers are the most common, with elements becoming steadily less abundant as the atomic number goes up. But that is only half the story.

EARLY DIP

Lithium, beryllium and boron are much less common than the general trend would suggest. This is because once hydrogen has fused into helium, the pair then find it easier to fuse into larger elements such as carbon. The three elements left in the middle were formed in large amounts by the effects of free protons zinging around in the early Universe. However, once the first stars came along, they consumed most of the lithium, beryllium and boron, making them relatively rare.

IN WITH A BANG

Both iron and nickel are bucking the trend. They have even atomic numbers but are still more abundant than one might expect. This is because elements of this size – roughly a third of the way through the table – are produced in huge amounts inside supernovae.

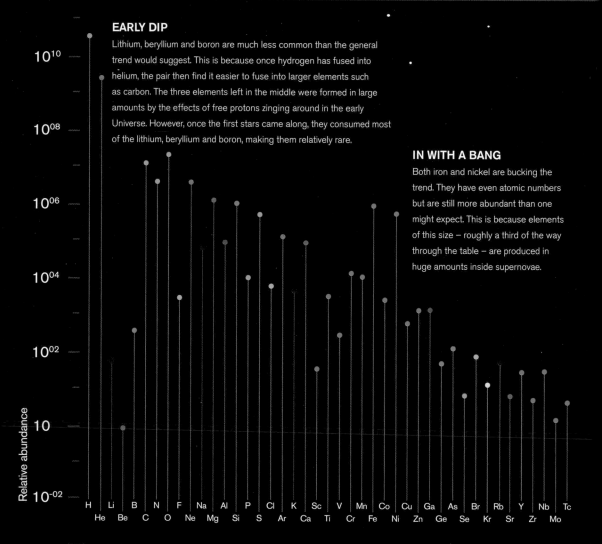

Relative abundance

10^{12}
10^{10}
10^{08}
10^{06}
10^{04}
10^{02}
10
10^{-02}

H | Li | B | N | F | Na | Al | P | Cl | K | Sc | V | Mn | Co | Cu | Ga | As | Br | Rb | Y | Nb | Tc
He | Be | C | O | Ne | Mg | Si | S | Ar | Ca | Ti | Cr | Fe | Ni | Zn | Ge | Se | Kr | Sr | Zr | Mo

EVENS WIN

This chart represents the universal abundance – how much is in the whole Universe – of the primordial elements. The trend of decreasing abundance for heavier elements is clear to see. However, elements with even atomic numbers are more common than those with odd atomic numbers. This is due to the Oddo–Harkins rule.

ODDO–HARKINS

The Oddo–Harkins rule predicts that elements with odd atomic numbers are generally less abundant than the elements either side with even atomic numbers. This is because protons in a nucleus are most stable when they are in pairs. Even atomic numbers represent nuclei made up of paired protons, while odd atomic numbers always include a lone proton. It is more likely that this straggler will be ejected from the nulceus, or pick up an extra partner during nuclear reactions inside stars. Thus, the even elements become more common than the odd ones.

LEAD

Lead is one of the heaviest elements with a large atomic number. Nevertheless, it is more common than the preceding 25 elements. This is because lead is the end point of most radioactive decay chains. Uranium and thorium break down through a long line of highly unstable elements until they become lead, which is stable. Over the history of the Universe, lead's abundance has been steadily increasing.

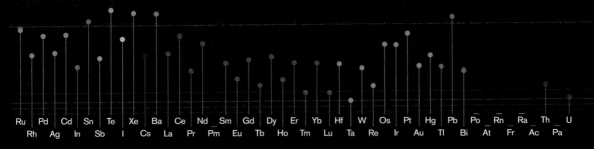

| Ru | Pd | Cd | Sn | Te | Xe | Ba | Ce | Nd | Sm | Gd | Dy | Er | Yb | Hf | W | Os | Pt | Hg | Pb | Po | Rn | Ra | Th | U |
| Rh | Ag | In | Sb | I | Cs | La | Pr | Pm | Eu | Tb | Ho | Tm | Lu | Ta | Re | Ir | Au | Tl | Bi | At | Fr | Ac | Pa | |

3

INSIDE CHEMISTRY

STATES OF MATTER

All elements and compounds have a standard state of matter – either solid, liquid or gas. Each one changes state at definitive temperatures, known as the melting and boiling points. These temperatures depend on how much energy it takes to make or break the bonds between atoms, which varies considerably.

PHYSICAL CHANGE

A change of state is a physical change, and does not alter the chemical behaviour of a substance. So steam reacts to form the same compounds as water does. However, steam holds more energy and so will react more quickly than water.

GAS

In the gas state, there are no bonds between atoms or molecules. All gas particles are free to move in any direction and so can take on any shape, and spread out to fill any volume.

Ionization

Recombination

Gas

Vaporization

Condensation

Liquid

Deposition

Sublimation

Melting

Freezing

Solid

LIQUID

In the liquid state, about ten per cent of the bonds seen in a solid are broken, which allows the atoms and molecules to move around and slide past each other. A liquid has a fixed volume but it can flow, changing its shape to fit any container.

SOLID

In the solid state, every atom is bonded to its neighbours. As a result, a solid has a fixed shape and volume.

PLASMA

Known as the fourth state of matter, plasma forms when energy (usually heat or electricity) is added to a gas. Molecules break up and individual atoms begin to shed electrons, forming an electrically charged substance. (The Sun is mostly made of plasma.) Forming plasma alters the atomic structure of an element, and thus it has a different set of physical and chemical properties.

TEMPERATURE SCALES

Temperature is an average measure of how much heat energy all the atoms in a substance contain. Temperature scales work by choosing a zero point and an upper point and then dividing their difference into degrees.

100° Celsius is the temperature at which pure water boils into steam.

The upper point of the Fahrenheit scale is based on human body temperature.

0° Celsius is the temperature at which pure water freezes into ice.

0° Fahrenheit was defined by using a 'frigorific mixture' of salts, which freeze when mixed.

The Kelvin scale uses the same degrees as Celsius. However, zero is set at the point at which atoms are unable to hold any thermal energy at all. This temperature (−273.15°C) is defined as absolute zero. It is impossible to make something this cold (although we can get very close).

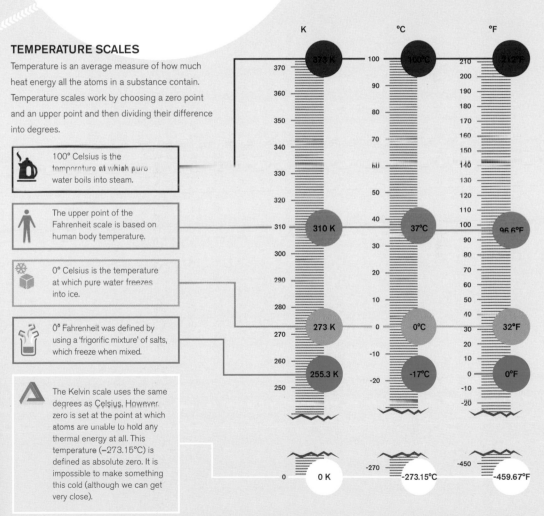

States of Matter | **81**

METALLIC BONDS

Of the 118 elements making up the periodic table, 84 are metals. Metals are generally shiny and hard. They conduct heat and electricity well and can be rolled flat and pulled into wires. These behaviours stem from the way that metal atoms are bonded together.

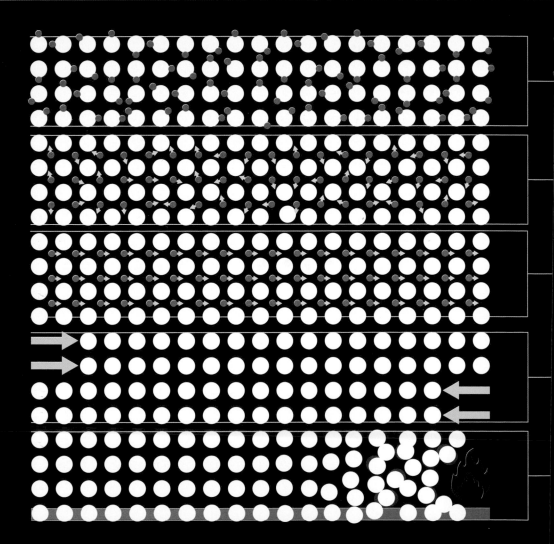

HEATING METALS

Since ancient times, metalworkers have been able to judge the temperature of a metal by its colour. This chart shows the colours typical of iron and steel. The coloured light is emitted by atoms, and the colour changes as the energy contained in these atoms increases. So in effect, the colour shows how strong the metallic bonds are.

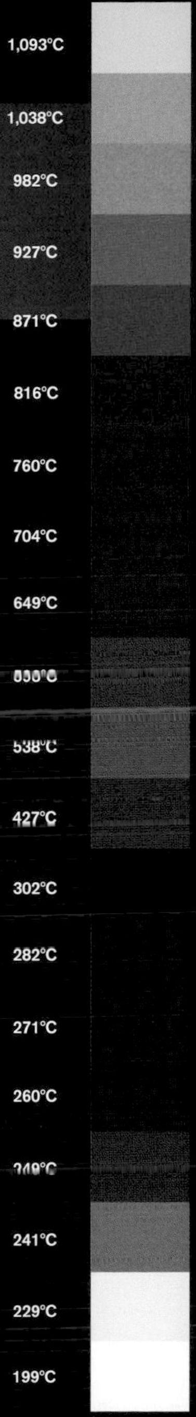

FREE ELECTRONS

An element has metallic properties because its atoms have few outer electrons. Most metals have one or two of them, although a few have more. This means that the outer shells of metal atoms are largely empty, and so the atoms readily shed their few outer electrons. This supply of free electrons underlies many metallic properties.

DELOCALIZED

The classical view of an atom depicts it with a fixed number of electrons associated with it. However, in a solid metal, the outer electrons of all atoms are delocalized and shared equally by all. This creates a 'sea' of charge around the atoms, gluing them together.

ELECTRICAL CONDUCTION

If a charge difference is placed across the metal – so it is more positive at one end, for example – the delocalized electrons (which are negative) will flow towards that end to rebalance the charge difference. This is the basis of an electric current.

MALLEABLE AND DUCTILE

The bonds formed by delocalized electrons hold metal atoms together strongly but not rigidly. The atoms can slide past each other without breaking apart. This is why metals are malleable (can be flattened) and ductile (can be pulled into wires).

HEAT CONDUCTION

Heat moves through metals more quickly than most non-metals. This is due to the metal atoms being able to move more freely. Adding heat at one end makes the atoms there vibrate more quickly. This motion is steadily transferred to neighbouring atoms so that the heat is conducted through the metal.

1,093°C
1,038°C
982°C
927°C
871°C
816°C
760°C
704°C
649°C
593°C
538°C
427°C
302°C
282°C
271°C
260°C
249°C
241°C
229°C
199°C

IONIC BONDS

When atoms of two or more elements bond together, they form a compound, which will invariably have markedly different properties to the original ingredients. Compounds formed from metal and non-metal elements generally employ ionic bonds.

MOVING ELECTRONS

A bond forms between two atoms because it puts their electrons into a more stable state that requires less energy. Ionic bonds do this by transferring outer electrons. Metal atoms generally have one or two outer electrons, and getting rid of those will make the atom stable. Non-metals do the opposite; they add electrons to their outer shell to increase their stability. In this example, the outer electron of a sodium atom has moved to a chlorine atom. Adding or losing electrons turns the atoms into an ion, an atom-like particle that carries electrical charge. A sodium ion has a charge of +1, and the chloride ion is −1. The opposite charges of the two ions pull them together, making an electrically neutral molecule of sodium chloride (NaCl; otherwise known as common salt).

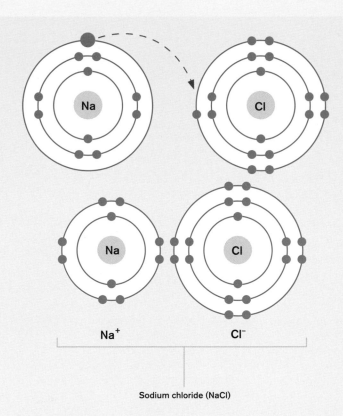

Sodium chloride (NaCl)

STAYING NEUTRAL

The charges involved in an ionic bond must add up to zero to create a neutral molecule. For example, when sodium reacts with oxygen to make sodium oxide (Na_2O, the stuff in glass), two sodium ions bond to a single oxide ion. Oxygen has two spaces to fill in its outer electrons, and so its ions have a charge of −2.

SHRINKING CATIONS

Atoms that lose electrons form positively charged ions, or cations. Cations have lost an entire electron shell, and so are considerably smaller than the atoms.

BULGING ANIONS

Negatively charged ions are called anions. They have filled their outer electron shells, so with the extra particles present, the negatively charged electrons spread out, making anions larger than their atoms.

 Li — Li⁺

 Be — 2+

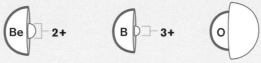 B — 3+

 O — 2−

 F — F⁻

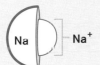 Na — Na⁺

 Mg — 2+

 Al — 3+

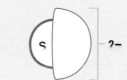 S — ?−

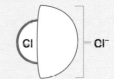 Cl — Cl⁻

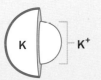 K — K⁺

 Ca — 2+

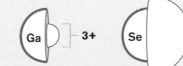 Ga — 3+

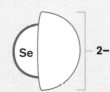 Se — 2−

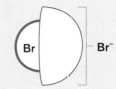 Br — Br⁻

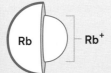 Rb — Rb⁺

 Sr — 2+

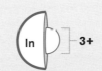 In — 3+

 Te — 2−

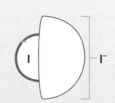 I — I⁻

COVALENT BONDS

Compounds formed from non-metals generally bond together using covalent bonds. Instead of transferring electrons between atoms, these bonds share them. This brings together their outer shells (also known as valence shells), creating a molecule.

MAKING EIGHT

The majority of valence shells have space for eight electrons. (Only hydrogen and helium have a two-electron outer shell.) An atom, such as chlorine, needs to make only one electron pair to form a stable molecule. However, oxygen, which has two free spaces, must find two electrons to pair with its own – as it does with hydrogen, to make water. Nitrogen makes three pairs, while carbon makes four pairs.

HOLDING ON

Non-metal atoms have mostly full valence shells, meaning that they hold on to their outer electrons with great force, because losing them makes the atoms less stable, not more so. As a result, most covalently bonded compounds will not conduct electricity, because there are very few free electrons to carry the charge.

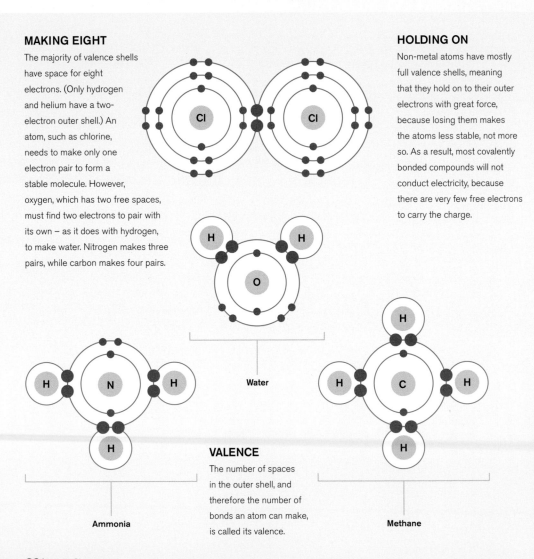

Water

Ammonia

Methane

VALENCE

The number of spaces in the outer shell, and therefore the number of bonds an atom can make, is called its valence.

MUTUAL REPULSION

The shared electron pairs in a covalent bond exert a repulsion on each other. This pushes them as far apart as possible, and gives each molecule a particular shape.

LONE PAIRS

The shape of a molecule is also defined by non-shared pairs of electrons that were there before the bonds formed. These pairs push each other away, and the shared pairs as well. Molecules comprising very different elements can take on similar shapes due to the action of their lone pairs. We show them here to represent their action, but lone pairs do not extend from the molecule like bonded atoms do.

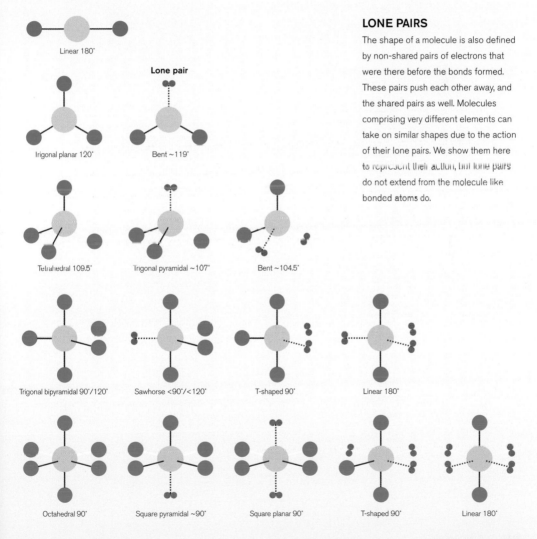

Linear 180°

Lone pair

Trigonal planar 120°

Bent ~119°

Tetrahedral 109.5°

Trigonal pyramidal ~107°

Bent ~104.5°

Trigonal bipyramidal 90°/120°

Sawhorse <90°/<120°

T-shaped 90°

Linear 180°

Octahedral 90°

Square pyramidal ~90°

Square planar 90°

T-shaped 90°

Linear 180°

REACTIONS

3

A chemical reaction is when reactants, comprising one or more element or compound, are converted into new substances, known as the products. During the reaction, chemical bonds are broken and reformed, resulting in products that are more stable than the reactants.

ACTIVATION ENERGY

Every reaction has an energy barrier to overcome before they start. This 'activation energy' is normally supplied by heating the reactants.

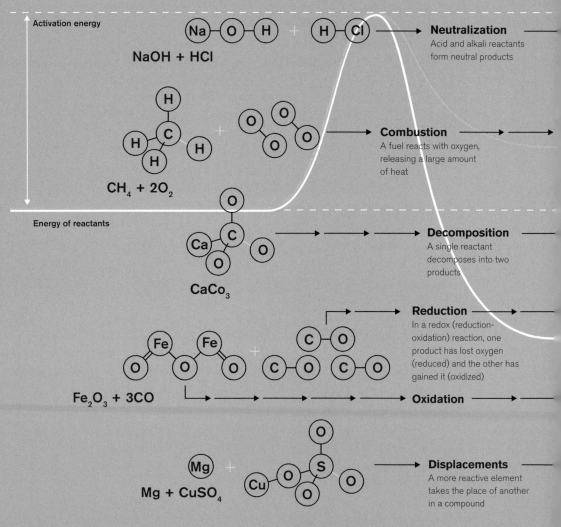

Activation energy

NaOH + HCl

Neutralization
Acid and alkali reactants form neutral products

CH$_4$ + 2O$_2$

Combustion
A fuel reacts with oxygen, releasing a large amount of heat

Energy of reactants

CaCo$_3$

Decomposition
A single reactant decomposes into two products

Fe$_2$O$_3$ + 3CO

Reduction
In a redox (reduction-oxidation) reaction, one product has lost oxygen (reduced) and the other has gained it (oxidized)

Oxidation

Mg + CuSO$_4$

Displacements
A more reactive element takes the place of another in a compound

ENERGY OUT

Breaking a chemical bond during a reaction releases energy, which is used to make new bonds in the products. If the new bonds require less energy than was released, the reaction gives out the spare heat – it is exothermic. However, some reactions are endothermic. They give out less heat than was put in, and so the products are very cold.

CATALYSTS

A catalyst is a substance that takes part in a reaction but is not used up by it. Its role is to reduce the activation energy so that the reaction occurs more easily.

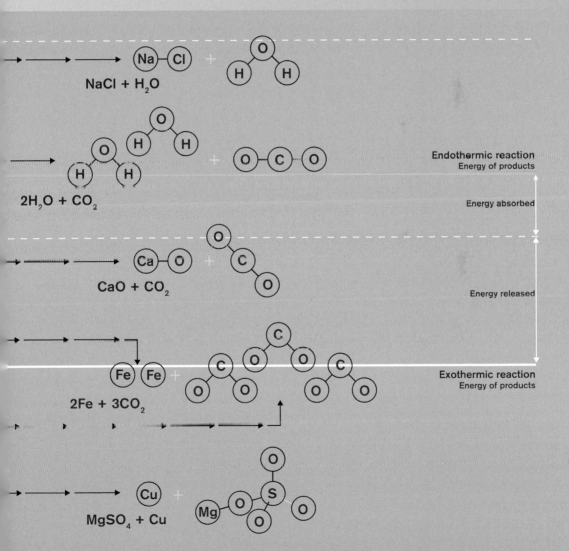

NaCl + H$_2$O

2H$_2$O + CO$_2$

Endothermic reaction
Energy of products

Energy absorbed

CaO + CO$_2$

Energy released

2Fe + 3CO$_2$

Exothermic reaction
Energy of products

MgSO$_4$ + Cu

MIXTURES

A great many everyday materials are mixtures. However, unlike a compound, which can only be broken into its constituents using chemical reactions, there are no chemical bonds between a mixture's constituents, so they can be separated using purely physical processes.

UNEVEN MIXTURES

The simplest mixtures involve substances that are unevenly dispersed, so each constituent remains easily identifiable. A mixture of coins is an everyday example of these so-called heterogeneous mixtures.

SEPARATING MIXTURES

There are a number of methods for separating mixtures. A dissolved solid can be removed by evaporating away the liquid part of the mixture, while large solids can be sorted from smaller ones using filters. Separating mixed liquids requires distillation, where the liquid with the lower boiling point is evaporated away and then cooled again so it condenses as a pure liquid.

EVEN MIXTURES

Homogeneous mixtures are those where the constituents are well mixed and individual ingredients are not visible. In these cases, one of the ingredients is the medium through which the others are dispersed. All states of matter can be mixed. There are three main types of homogeneous mixture: suspensions, colloids (or emulsions) and solutions.

Matter

Can it be physically separated?

No

Yes

Pure substance

Mixture

Can it be chemically decomposed?

Is the composition the same throughout?

No

Yes

No

Yes

Element

Compound

Emulsion or suspension

Solution

FOAM
Beaten egg white,
shaving cream,
whipped cream,
ice-cream soda

 Dispersed particles
Gas

 Dispersion medium
Liquid

 Dispersed particles
Gas

 Dispersion medium
Solid

SOLID FOAM
Marshmallow,
rubber foam

SUSPENSION
The dispersed ingredients are very large compared to the medium and will settle, given time. Muddy water is a suspension.

LIQUID AEROSOLS
Clouds, fog,
mist, hairspray,
deodorant spray

 Dispersed particles
Liquid

 Dispersion medium
Gas

COLLOIDAL SOLUTION
The dispersed constituents are very small, but still larger than the molecules of the medium. Shampoo is a colloidal solution.

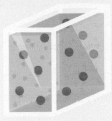

 Dispersed particles
Liquid

 Dispersion medium
Liquid

EMULSION
Milk,
mayonnaise,
blood

SOLUTION
The dispersed ingredients are dissolved, which means their molecules are as evenly spread as the molecules of the medium itself. Saltwater is a solution.

GEL
Cheese,
butter,
margarine

 Dispersed particles
Liquid

 Dispersion medium
Solid

RADIOACTIVITY

All elements have radioactive forms, or isotopes, and there are 38 elements that have no stable isotopes at all. Radioactivity occurs when an atom's nucleus is unstable, so it breaks apart, or decays, releasing high-energy particles.

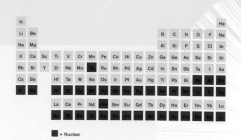

■ = Nuclear

ATOMIC FISSION

This unusual form of radioactive decay occurs when a nucleus splits into two roughly equal parts, releasing a huge amount of energy. If the fission event creates the conditions for further fission, then a chain reaction occurs. If uncontrolled, that makes a nuclear explosion. Nuclear reactors control the reaction to create a power source. The main fissile isotope is uranium-238, which splits into krypton and barium atoms.

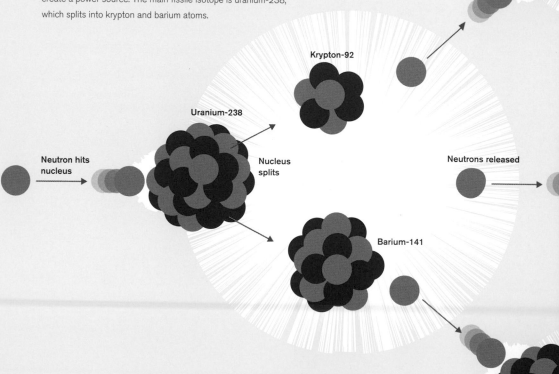

Krypton-92

Uranium-238

Neutron hits
nucleus

Nucleus
splits

Neutrons released

Barium-141

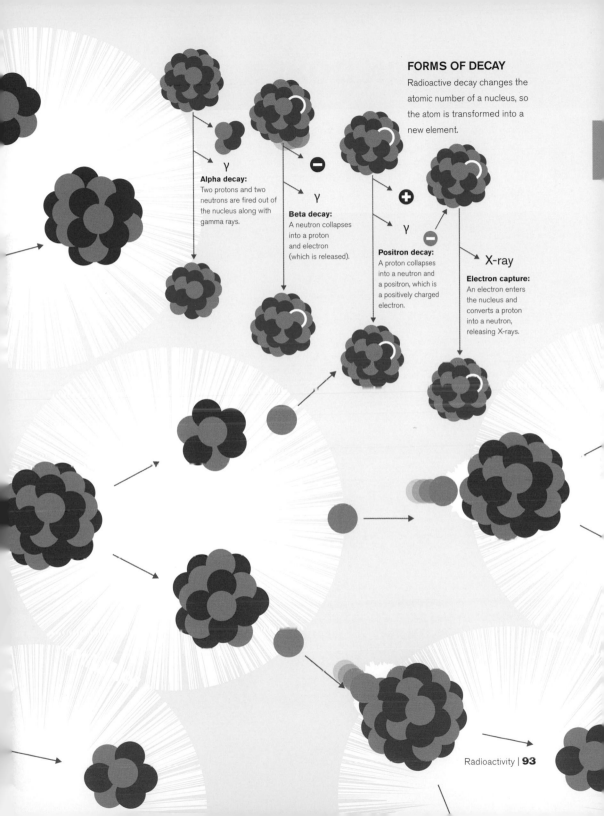

FORMS OF DECAY

Radioactive decay changes the atomic number of a nucleus, so the atom is transformed into a new element.

γ

Alpha decay:
Two protons and two neutrons are fired out of the nucleus along with gamma rays.

γ

Beta decay:
A neutron collapses into a proton and electron (which is released).

γ

Positron decay:
A proton collapses into a neutron and a positron, which is a positively charged electron.

X-ray

Electron capture:
An electron enters the nucleus and converts a proton into a neutron, releasing X-rays.

RADIATION DOSES

Radioactivity is entirely natural. There are radioactive isotopes in rocks, the atmosphere – even in food and the body. These create a low level of background radiation. However, the refined radioactive materials used in power production, medicine and weapons add to exposure and must be monitored.

EXPOSURE

Exposure to radioactivity is measured in sieverts (Sv), which shows how much energy from radioactivity is in each kilogram of body tissue. A millisievert (10^{-3} sieverts) is written as mSv, and a microsievert (10^{-6} sieverts) is written as µSv.

● = 0.05 µSv ● = 0.02 mSv ● = 10mSv

Sleeping next to someone (0.05 µSv)

Living within 80km of a nuclear power plant for a year (0.09 µSv)

Eating one banana (0.1 µSv)

Living within 80km of a coal power plant for a year (0.3 µSv)

Dental or hand X-ray (5 µSv)

Background dose received by an average person over one normal day (10 µSv)

Aeroplane flight from New York to LA (40 µSv)

Living in a stone, brick, or concrete building for a year (70 µSv)

EPA yearly release limit for a nuclear power plant (250 µSv)

Yearly dose from natural potassium in the body (390 µSv)

Normal yearly background dose. About 85% is from natural sources. Nearly all of the rest is from medical scans (~3.65 mSv)

CUMULATIVE EFFECTS

Radioactive substances accumulate in the body. Therefore many of the doses in this chart relate to exposure over a certain time period.

BIOLOGICAL EFFECTS

The energy from radioactive decay inside the body can alter (or denature) the many intricate chemicals used in metabolism. The body is able to identify and remove these damaged chemicals, but there is a limit to its ability to do that. High exposure to radioactivity increases the likelihood of cancers and causes radiation sickness. This has a body-wide effect that has a disastrous impact on body tissues that are constantly growing and renewing their cells, such as the lining of the stomach, the skin, the blood supply and the genitals. Treatment involves using chemicals to capture and remove radioactive substances.

Maximum yearly dose permitted for US radiation workers (50 mSv)

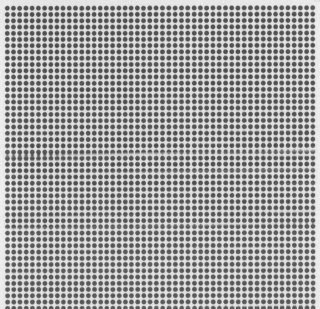

Lowest one-year dose clearly linked to increased cancer risk (100 mSv)

Fatal dose even with treatment (8 Sv)

Dose limit for emergency workers in lifesaving operations (250 mSv)

STABILITY

3

Every atomic nucleus has a half-life. This is the time it takes for half of a sample of nuclei to decay. Highly radioactive nuclei have half-lives of millionths of a second and decay away very fast. The half-lives of non-radioactive elements are measured in trillions of years.

STABLE NUCLEI

Although there is a theoretical half-life for all nuclei, the half-lives of non-radioactive elements far outstrip the age of the Universe.

120 —

PROTON-NEUTRON RATIO

This chart plots the ratio of protons (Z) to neutrons (N) in all known atomic nuclei. The colours represent the half-lives of each nuclei. The stable nuclei, those of the atoms that make up the Earth and Universe around us, form a dark band in the centre.

100 —

INCREASING RATIO

In the smaller elements, the number of neutrons and protons is roughly equal, but as nuclei get steadily heavier, the ratio begins to skew towards the neutrons. This is because inside larger nuclei, protons are further apart and more likely to push each other away. To maintain stability, the protons are diluted with electrically neutral neutrons.

80 —

60 —

40 —

20 —

N

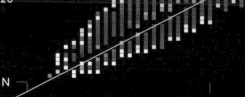

Z ⟍ 20 ⟍ 40 ⟍ 60

N (NUMBER OF NEUTRONS)

Z (NUMBER OF PROTONS)

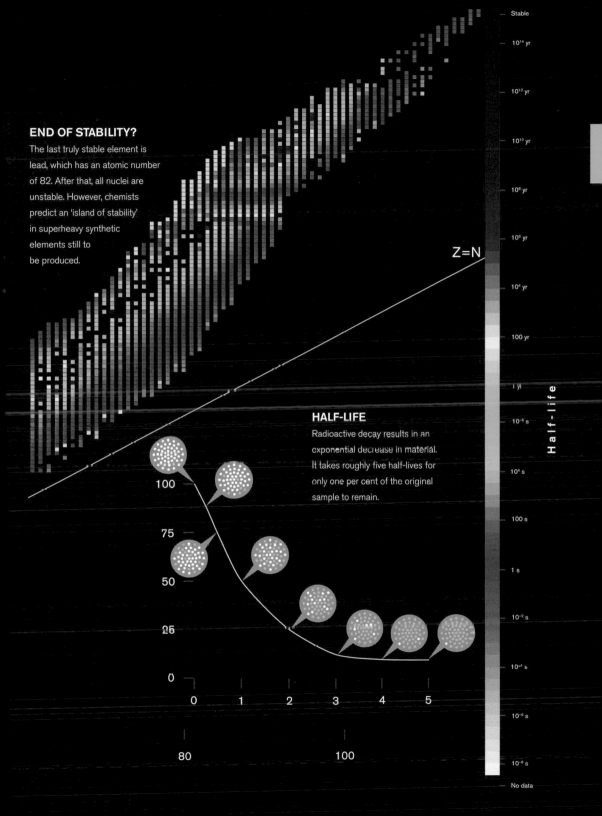

END OF STABILITY?

The last truly stable element is lead, which has an atomic number of 82. After that, all nuclei are unstable. However, chemists predict an 'island of stability' in superheavy synthetic elements still to be produced.

Z=N

HALF-LIFE

Radioactive decay results in an exponential decrease in material. It takes roughly five half-lives for only one per cent of the original sample to remain.

100

75

50

25

0

0   1   2   3   4   5

80   100

Stable

10^{14} yr

10^{12} yr

10^{10} yr

10^{8} yr

10^{6} yr

10^{4} yr

100 yr

1 yr

10^{-6} s

10^{4} s

100 s

1 s

10^{-2} s

10^{-4} s

10^{-6} s

10^{-8} s

No data

Half-life

HOW TO MAKE A NEW ELEMENT

3

The radioactive metal uranium is the heaviest atom to exist in appreciable quantities in nature. However, since the 1930s, scientists have been enlarging the periodic table by manufacturing new elements.

TRANSURANIUM

The majority of synthetic elements make up the so-called transuranium group, because they are all heavier than uranium.

ATOM SMASHING

Making synthetic elements is a mixture of precision and brute force. Simply put, a stream of smaller nuclei (A) – atoms that have had their electrons removed – is fired at a target made of larger nuclei (B). Most of the time the impacts have no effect, but a few will be perfectly aligned so that a large nucleus captures a smaller one, and the pair merge into the nuclei of a yet larger – and completely new – element (C).

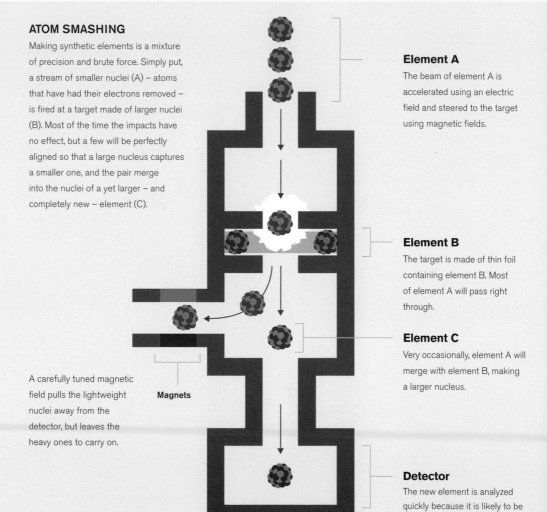

Element A

The beam of element A is accelerated using an electric field and steered to the target using magnetic fields.

Element B

The target is made of thin foil containing element B. Most of element A will pass right through.

Element C

Very occasionally, element A will merge with element B, making a larger nucleus.

A carefully tuned magnetic field pulls the lightweight nuclei away from the detector, but leaves the heavy ones to carry on.

Magnets

Detector

The new element is analyzed quickly because it is likely to be a very unstable isotope.

BOMB BORN

Many of the first synthetic elements, such as einsteinium, were made as a by-product of the enormous explosions of nuclear weapons tests.

INGREDIENTS LIST

This chart shows what smaller elements are smashed together to build the nuclei of larger, superheavy synthetic elements. Often scientists use synthetic nuclei to build even heavier ones.

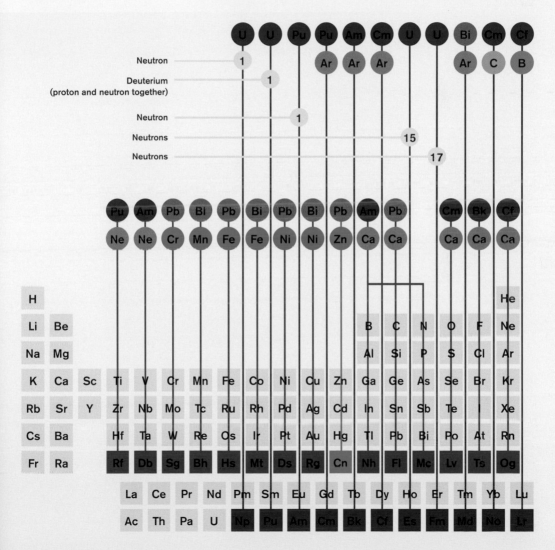

ORGANIC CHEMISTRY

3

About 90 per cent of the ten million compounds investigated by chemists to date contain carbon. A carbon atom is able to form four bonds at once, and so its compounds are highly varied. The chemistry of carbon is termed 'organic', because many of its compounds were made by or derived from living things.

HYDROCARBONS

The most simple class of organic compounds are the hydrocarbons. These are compounds with hydrogen, many of which are used as fuels.

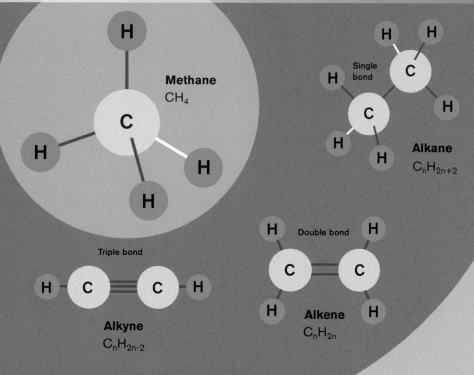

Methane
CH_4

Single bond

Alkane
C_nH_{2n+2}

Triple bond

Double bond

Alkyne
C_nH_{2n-2}

Alkene
C_nH_{2n}

NAMING CONVENTION

The names of organic compounds follow a system. The suffix, such as '-ane', '-ene' or '-yne', tells you what kind of compound it is. The prefix denotes the number of carbons chained together.

1	Meth	Methane	CH_2
2	Eth	Ethane	C_2H_6
3	Prop	Propane	C_3H_8
4	But	Butane	C_4H_{10}
5	Pent	Pentane	C_5H_{12}
6	Hex	Hexane	C_6H_{14}
7	Hept	Heptane	C_7H_{16}
8	Oct	Octane	C_8H_{18}
9	Non	Nonane	C_9H_{20}
10	Dec	Decane	$C_{10}H_{22}$

ISOMERS

Organic compounds can arrange the same number of atoms in different ways. This creates a family of compounds called isomers.

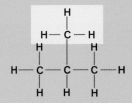

C_4H_{10}

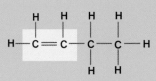

HANDEDNESS

Isomers can have handedness (also known as chirality). These two isomers are mirror images of each other.

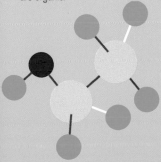

CIS-TRANS

The orientation of sections of the molecule is also important in isomerism. A trans-isomer has them on opposite sides, while they are on the same side in a cis-isomer.

C_4H_8

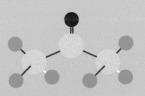

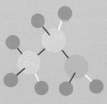

COMMON CHEMICALS

Many of the most familiar chemicals are organic.

The molecules of alcohols, such as ethanol, contain an oxygen-hydrogen grouping.

Aldehydes, such as formaldehyde, are used as preservatives.

Ketones are used as solvents.

Thiols, which contain sulphur, make pungent smells.

Amines, which contain nitrogen, make 'meaty' smells.

pH CHART

The strength of acids and alkalis is measured on the pH scale, which runs from 0 to 14. A pH of 7 is neutral (neither acidic nor alkaline); acids are less than 7; while alkalis have a pH higher than 7. The scale is logarithmic, which means a difference of one in the scale relates to a ten-fold increase or decrease in strength.

WHY ACID?

pH stands for 'potential hydrogen'. Acids are chemicals that contribute hydrogen ions (H⁺) to reactions. Stronger acids deliver more ions and therefore react more violently.

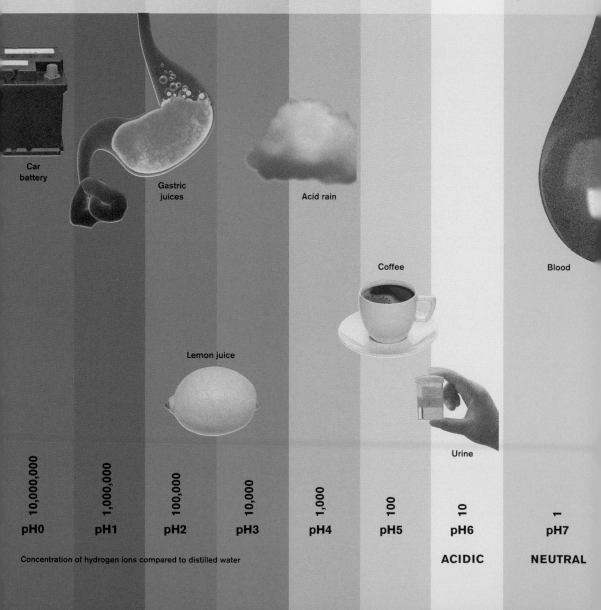

Car battery

Gastric juices

Acid rain

Coffee

Blood

Lemon juice

Urine

10,000,000	1,000,000	100,000	10,000	1,000	100	10	1
pH0	**pH1**	**pH2**	**pH3**	**pH4**	**pH5**	**pH6**	**pH7**
						ACIDIC	**NEUTRAL**

Concentration of hydrogen ions compared to distilled water

WHY ALKALI?

An alkali is the opposite of an acid. It brings hydroxide ions (OH⁻) to reactions. These react with hydrogen ions to make water. Therefore when acids and alkalis react, one of the products is always water.

INDICATORS

Coloured chemicals are used to indicate the pH of a solution. These colours are based on the universal indicator.

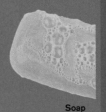

Indigestion medicine

Baking soda

Soap

Ammonia

Bleach

Drain cleaner

1/10	1/100	1/1,000	1/10,000	1/100,000	1/1,000,000	1/10,000,000
pH8	**pH9**	**pH10**	**pH11**	**pH12**	**pH13**	**pH14**

ALKALINE

Concentration of hydrogen ions compared to distilled water

CHEMISTRY OF GEMS

Gems are highly prized crystals that share a certain set of properties. They are invariably hard, and therefore not damaged easily by knocks and scrapes, and they are transparent to light, meaning that when they are cut in the right ways, they sparkle with an inner fire. But what really sets gems apart is their colour.

BEING REFLECTIVE

A gem's colour arises because that is the only light that is reflected by the stone. All other light is absorbed by its crystal lattice.

Sapphire

Compound: Aluminium oxide
Impurity: Titanium

Diamond

Pure carbon

Turquoise

Compound: Aluminium hydroxide
Impurity: Copper

Jade

Compound: Sodium aluminium silicate
Impurity: Chromium and iron

Peridot

Compound: Magnesium silicate
Impurity: Iron

Garnet

Compound: Magnesium aluminum silicate
Impurity: Iron

Amethyst

Compound: Silicon dioxide
Impurity: Iron

Citrine

Compound: Silicon dioxide
Impurity: Aluminium

LATTICE STRUCTURE

A gem's lattice structure locks all of
its atoms into a rigid, repeating network.
This is where the stone's hardness comes from.

CRUCIAL IMPURITIES

As this chart shows, gems are generally similar (or even
identical) mineral compounds, many of which are normally
colourless. The gem's striking colours come from the
inclusion of tiny amounts of impurities.

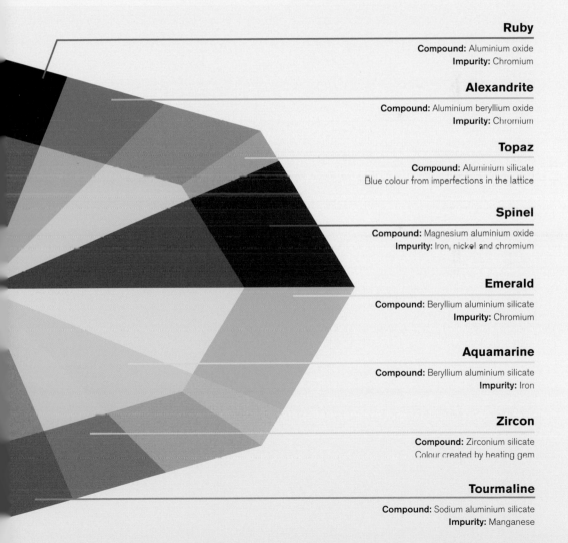

Ruby

Compound: Aluminium oxide
Impurity: Chromium

Alexandrite

Compound: Aluminium beryllium oxide
Impurity: Chromium

Topaz

Compound: Aluminium silicate
Blue colour from imperfections in the lattice

Spinel

Compound: Magnesium aluminium oxide
Impurity: Iron, nickel and chromium

Emerald

Compound: Beryllium aluminium silicate
Impurity: Chromium

Aquamarine

Compound: Beryllium aluminium silicate
Impurity: Iron

Zircon

Compound: Zirconium silicate
Colour created by heating gem

Tourmaline

Compound: Sodium aluminium silicate
Impurity: Manganese

4

DIRECTORY OF ELEMENTS

HYDROGEN

H	1

Hydrogen is the first element in the periodic table. It has the simplest atom of all, being made up of a single proton nucleus with a single electron held in orbit around it. Hydrogen is in Group 1 of the table, but unlike other members, it is not a metal but an ultra-lightweight gas. Compounds containing hydrogen ions are known as acids.

VISIBLE UNIVERSE

The elements make up the visible matter – the stars, planets and galaxies – that we see in the Universe around us. Three-quarters of that is made from hydrogen. All the other 92 or so primordial elements make up the remaining quarter.

25%

HYDROGEN MATTERS

Hydrogen is the most abundant element in the Universe because it was the first one to be formed. Even so, it took 380,000 years after the Big Bang for the Universe to cool sufficiently for hydrogen atoms to form.

GOING DARK

In the 1930s, astronomers discovered 'dark matter', which is invisible and not made from elements. In the late 1990s they discovered 'dark energy', a mysterious anti-gravity effect that is pushing space apart. Together, dark matter and dark energy make up

75%

96%

of the Universe. For all its abundance, hydrogen comprises just three per cent and the other elements add up to one per cent.

Atomic weight: 1.00794
Colour: n/a
Phase: gas
Melting point: -259°C (-434°F)
Boiling point: -253°C (-423°F)
Crystal structure: n/a

Category: non-metal
Atomic number: 1

^1H ^1H ^1H ^1H

4%

FUSION POWER

Like most stars, the Sun is a ball of hydrogen plasma that is collapsing under its own gravity. The huge pressure at the solar core is so powerful that it fuses hydrogen atoms together. First two hydrogen atoms make an atom of deuterium (^2H), which is an isotope of hydrogen that has a neutron and proton in its nucleus. Next a hydrogen and deuterium fuse making tritium (^3H), an isotope with two neutrons. Finally, two tritium fuse, forming an atom of helium (^4He), the next heaviest element.

^2H ^1H ^1H ^2H

γ   γ

^3H ^3H

Nuclear fusion releases the heat and light that lights up the star. In the process, four per cent of the atoms' mass is converted into pure energy. The Sun is slowly eating itself through fusion.

^1H ^1H

^4He

HELIUM

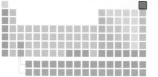

He 2

Atomic weight: 4.002602
Colour: n/a
Phase: gas
Melting point: -272°C (-458°F)
Boiling point: -269°C (-452°F)
Crystal structure: n/a

Category: noble gas
Atomic number: 2

Helium is the first member of the noble gas group, so called because they are chemically inert and will not combine with 'common' elements. Helium was the first noble gas to be discovered and it was found in an unusual place – light from the Sun. During a solar eclipse in 1868, astronomers were studying the corona, a ring of glowing gas that surrounds the Sun. They found a spectrum of coloured lights that did not match a known element. This showed that the Sun contained a new element, which was named helium, meaning 'sun metal'. In 1895, chemists found helium on Earth escaping from radioactive rocks and volcanoes – and it was found to be a lightweight gas.

Emission spectrum of helium

SQUEAKY

Helium has many important uses, and using it to make your voice squeaky is not one of them. The helium squeak is caused by the low-density gas vibrating more quickly than thicker air as it passes through the vocal cords.

Air

Helium

LITHIUM

Atomic weight: 6.941
Colour: silver–white
Phase: solid
Melting point: 181°C (358°F)
Boiling point: 1,342°C (2,448°F)
Crystal structure: body-centred cubic

Category: alkali metal
Atomic number: 3

Lithium is the first metal in the periodic table. It is used as a medicine for mood disorders, such as bipolar, and it is the trigger mechanism for thermonuclear weapons, or H-bombs. However, by far the biggest use of lithium is in compact, energy-dense rechargeable batteries – the kind of technology that powers our phones now and will drive the electric cars of tomorrow.

**Chile
12,900
tonnes**

**Australia
13,000 tonnes**

SALT MINING

Lithium ores are found in rocks, but the principal source is salt flats, especially in the Andes region. This production chart belies the fact that Bolivia is set to become the world's biggest lithium producer in the near future, with half the global reserves.

**China
5,000
tonnes**

**Argentina
2,900 tonnes**

**Zimbabwe
1,000 tonnes**

**Portugal
570 tonnes**

**Brazil
400 tonnes**

BERYLLIUM

Be 4

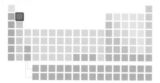

Atomic weight: 9.012182
Colour: silvery white
Phase: solid
Melting point: 1,287°C (2,349°F)
Boiling point: 2,469°C (4,476°F)
Crystal structure: hexagonal

Category: alkaline earth metal
Atomic number: 4

James Webb Space Telescope

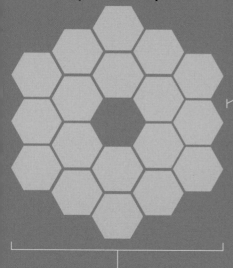

6.5m

Beryllium is an unreactive metal that has very high thermal stability. That means it does not expand or warp when it is heated.

13 million light years

HEAT MIRROR

Beryllium was used to make the mirror for the James Webb Space Telescope (JWST). With the largest mirror ever launched into space, this telescope is designed to pick up heat rays from space, not light.

Hubble

2.4m

12 million light years

STRETCHED LIGHT

The light from the oldest and most distant stars has been travelling through space for so long, it has been stretched into invisible heat waves. That means the JWST can see stars that are more than one million light years further away than the Hubble Space Telescope, which only picks up light.

BORON

Atomic weight: 10.8111
Colour: varied
Phase: solid
Melting point: 2,076°C (3,769°F)
Boiling point: 3,927°C (7,101°F)

Crystal structure: rhombohedral (as boride)
Category: metalloid
Atomic number: 5

5

B

Boron is a hard, dark soild with a dull shine. Despite its name, this element is far from boring, and has a diverse set of uses.

CONTROL RODS IN NUCLEAR REACTOR

The boron inside the rods absorbs neutrons produced by nuclear fissions. The rods are lowered into the reactor to slow down the chain reaction.

Heatproof glass

TELEVISION STONE

Ulexite, a boron mineral, has a unique optical property, where light from under the crystal is transmitted to the top, making a clear image.

Silly Putty

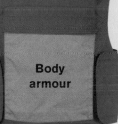

Body armour

CARBON

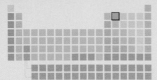

Atomic weight: 12.0107
Colour: clear (diamond) and black (graphite)
Phase: solid
Melting point: n/a – turns to vapour (sublimes) before it melts

Sublimation point: 3,642°C (6,588°F)
Crystal structure: hexagonal (graphite) and face-centred cubic (diamond)
Category: non-metal
Atomic number: 6

Carbon is the basis of all chemicals found in living things. The supply of this element moves through the biosphere via the carbon cycle.

GAS FLOW

Plants use carbon dioxide (CO_2) to make sugars by photosynthesis. This gas is returned to the air by respiration, a process used by plants, and the animals that eat them, to burn sugars and release energy.

FOSSIL SUPPLY

Dead organisms are mostly eaten by micro-organisms, which also release the carbon as CO_2. Some dead material becomes locked into rocks, and the carbon-rich substances form coal and petroleum oil and gas.

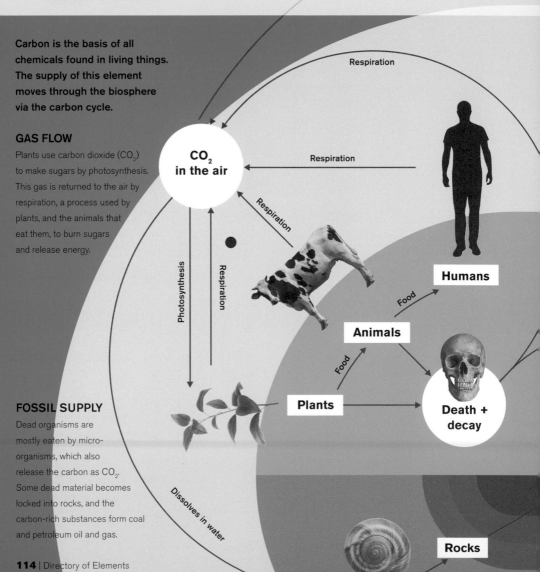

Respiration

CO_2 in the air

Respiration

Respiration

Photosynthesis

Respiration

Humans

Food

Animals

Food

Plants

Death + decay

Dissolves in water

Rocks

Shells

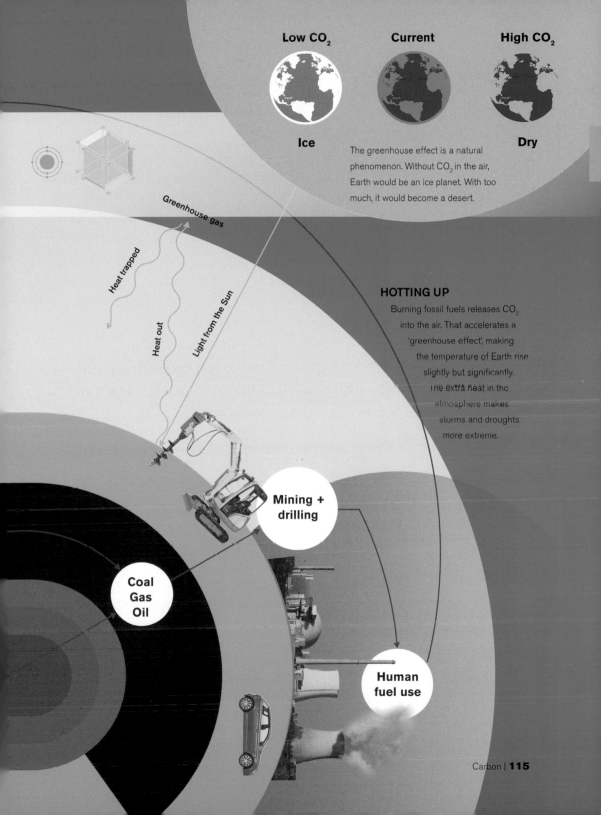

Low CO$_2$ **Current**   **High CO$_2$**

Ice   **Dry**

The greenhouse effect is a natural
phenomenon. Without CO$_2$ in the air,
Earth would be an ice planet. With too
much, it would become a desert.

Greenhouse gas

Heat trapped

Heat out

Light from the Sun

HOTTING UP

Burning fossil fuels releases CO$_2$
into the air. That accelerates a
'greenhouse effect', making
the temperature of Earth rise
slightly but significantly.
The extra heat in the
atmosphere makes
storms and droughts
more extreme.

**Mining +
drilling**

**Coal
Gas
Oil**

**Human
fuel use**

NITROGEN

N 7

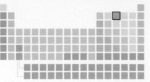

Nitrogen is an unreactive gas that makes up four-fifths of the air we breathe. Nevertheless, nitrogen is a crucial element used to build proteins, the work-horse chemicals of cells. Modern industrial agriculture must add nitrogen chemicals to grow enough food. A third of humanity eats food that uses an artificial nitrogen source.

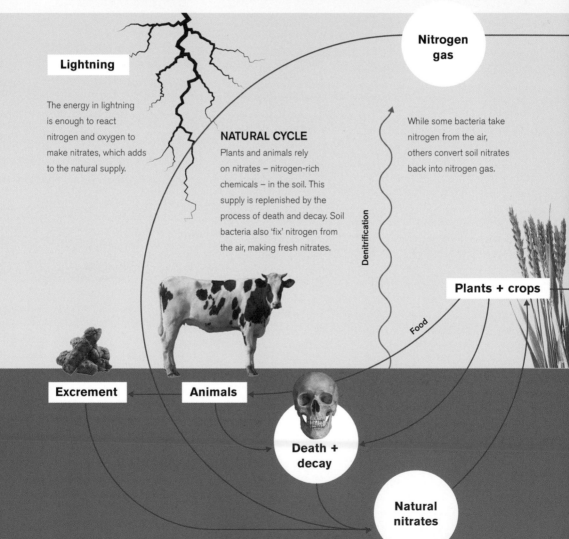

Lightning

The energy in lightning is enough to react nitrogen and oxygen to make nitrates, which adds to the natural supply.

NATURAL CYCLE

Plants and animals rely on nitrates – nitrogen-rich chemicals – in the soil. This supply is replenished by the process of death and decay. Soil bacteria also 'fix' nitrogen from the air, making fresh nitrates.

Nitrogen gas

While some bacteria take nitrogen from the air, others convert soil nitrates back into nitrogen gas.

Denitrification

Food

Plants + crops

Excrement

Animals

Death + decay

Natural nitrates

Atomic weight: 14.00672
Colour: none
Phase: gas
Melting point: -210°C (-346°F)
Boiling point: -196°C (-320°F)
Crystal structure: n/a

Category: non-metal
Atomic number: 7

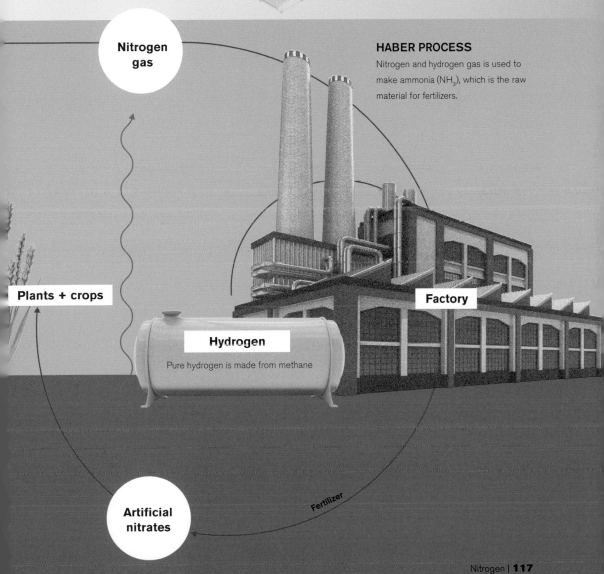

Nitrogen
gas

HABER PROCESS

Nitrogen and hydrogen gas is used to make ammonia (NH_3), which is the raw material for fertilizers.

Plants + crops

Hydrogen

Pure hydrogen is made from methane

Factory

Fertilizer

Artificial
nitrates

OXYGEN

O 8

Oxygen is a reactive gas. It is highly electronegative, which means it pulls strongly on electrons, attracting them to its outer shell to form bonds in chemical reactions. There are a few more reactive elements in the periodic table, but oxygen is unusual. Despite its readiness to form compounds, oxygen is found pure in nature.

HEAVY WATER

Seventy per cent of Earth's surface is covered in water, which is composed of hydrogen and oxygen. Despite the hydrogen atoms outnumbering oxygens by a factor of two, heavier oxygen makes up 88 per cent of the ocean's mass.

IN ABUNDANCE

Oxygen is the most common element in the Earth's crust. It makes up 49 per cent of the mass of Earth's rocks, being found in most minerals including silica (sand), clays and limestones.

INTO THIN AIR

The air pressure at the top of Mount Everest is a third of the standard pressure at sea level. That pressure is too low for oxygen to pass efficiently from the air to the blood. It is not sufficient to simply breathe three times faster, and as a result the region around the summit, or any mountain more than 8,000 metres high, is known as the Death Zone. Spending time there gradually degrades the body, and will eventually lead to a loss of consciousness and death.

TOXIC GAS

All of the pure oxygen in the air is released by Earth's plants and other photosynthetic life. No pure oxygen was released from rocks when the planet's atmosphere formed, and for the first two billion years of its history, Earth was swathed mostly in nitrogen and carbon dioxide. The earliest life forms did not require oxygen, getting their energy instead by using other chemicals, such as sulphur. Photosynthesis evolved 2.3 billion years ago, resulting in a huge release of oxygen. Oxygen was toxic to earlier life forms, and so photosynthesis would have resulted in the extinction of most of the life on Earth in a period known as the Great Oxygenation Event. Ironically, photosynthesis is the foundation of all food webs. Without it, animal life as we know it today would be impossible.

Atomic weight: 15.9994
Colour: colourless
Phase: gas
Melting point: -219°C (-362°F)
Boiling point: -183°C (-297°F)
Crystal structure: n/a

Category: non-metal
Atomic number: 8

MAGNETIC LIQUID

Oxygen is paramagnetic, which means it is attracted to magnetic fields. The interaction is so weak when oxygen is a gas that its effect is negligible. However, when oxygen is cooled into a liquid, the force gets stronger. A magnet can bend a stream of liquid oxygen.

MONSTERS

Oxygen makes up 21 per cent of the atmosphere, but in the distant past, there have been periods when there was more oxygen around. For example, in the Carboniferous period, about 300 million years ago, the first trees evolved and pushed out more of the gas, making the air mix rise to 35 per cent oxygen. This made it possible for invertebrates, which take in oxygen directly through the body surface, to grow to huge sizes. For example, meganeura, a dragonfly, grew to 65 centimetres long, while arthropleura, a relative of the millipede, reached 2.3 metres.

A NEW AIR

Pure oxygen was first isolated by the Swede Carl Scheele in 1772. He did not publicise his discovery of this 'fire air' widely, and so most of the credit goes to Englishman Joseph Priestley, who made pure oxygen in 1774. Priestley named it 'dephlogisticated air', a mouthful derived from a theory about the process of burning. This stated that material burned by absorbing a mysterious substance called 'phlogiston'. Oxygen fed fires so effectively that Priestley reasoned it contained no phlogiston – or it was dephlogisticated. Chemistry soon moved on, and Antoine Lavoisier renamed the gas oxygen.

FLUORINE

F 9

Atomic weight: 18.9984032
Colour: pale yellow
Phase: gas
Melting point: -220°C (-364°F)
Boiling point: -188°C (-307°F)
Crystal structure: n/a

Category: halogen
Atomic number: 9

Fluorine is the most reactive non-metal element. A pure jet of fluorine gas can burn through most materials, including brick and solid iron. Early attempts to make pure fluorine resulted in the gas destroying the apparatus. After 74 years of research by many chemists, Henri Moissan finally succeeded in 1884 by cooling his equipment to very low temperatures to slow the reactions.

DANGER

Pure fluorine must be carefully controlled to prevent dangerous reactions. It is stored as a super-cooled liquid (below −200°C/−392°F) in nickel or copper containers, which do not react strongly with fluorine.

Fluorine makes the 'F' in CFC, or chlorofluorocarbon. CFC gases were used in aerosol cans, but were banned because they damage the atmosphere.

Fluoride compounds are in toothpastes. The fluoride ions replace phosphate ions in tooth enamel, creating a stronger structure that can withstand attack by food acids.

A fluorine-based liquid can be used to deliver oxygen in the lungs, so people breathe liquid.

Teflon, a slippery plastic used in non-stick pans, contains fluorine.

NEON

1,340 parts
per million

Universe

Atomic weight: 20.1797
Colour: colourless
Phase: gas
Melting point: -249°C (-415°F)
Boiling point: -246°C (-411°F)
Crystal structure: n/a

Category: noble gas
Atomic number: 10

10
Ne

Neon is the fifth-most common element in the Universe. However, it is rare on Earth. Neon makes up less than 20 millionths of the atmosphere. However, this rare gas does have a famous use – neon lights.

WASTE GAS

Neon is a waste product of the distillation of air used to make pure oxygen and nitrogen. In this process, air is cooled to around -250°C so it becomes a liquid. The liquid is allowed to warm up, so each gas in the mixture can boil away. The first gases to be collected are tiny amounts of neon and the other noble gases.

Atmosphere — 18 parts per million

Rocks

30 parts per million

Air

Distillation

-150

-200

-250

Kr He Xe

Ar

Ne

N

O

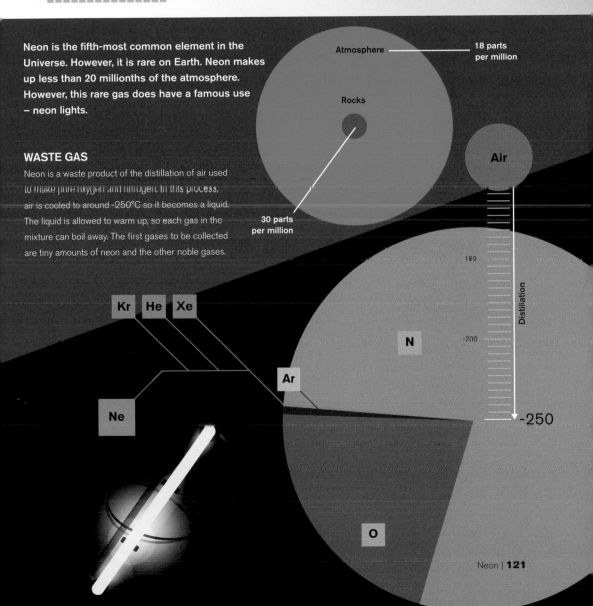

SODIUM

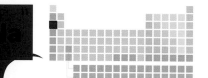

Atomic weight: 22.989770
Colour: silvery white
Phase: solid
Melting point: 98°C (208°F)
Boiling point: 883°C (1,621°F)
Crystal structure: body-centred cubic

Category: alkali metal
Atomic number: 11

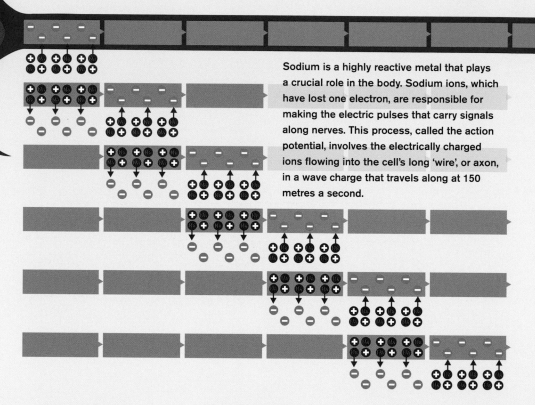

Sodium is a highly reactive metal that plays a crucial role in the body. Sodium ions, which have lost one electron, are responsible for making the electric pulses that carry signals along nerves. This process, called the action potential, involves the electrically charged ions flowing into the cell's long 'wire', or axon, in a wave charge that travels along at 150 metres a second.

SOAPY CHEMICAL

Sodium is one of the chemicals in soap. Sodium stearate is the white, greasy solid that makes the dirt mix with water.

SALTS NEEDED

Sodium chloride, better known as common salt, is the main source of sodium in food. Without enough salt, muscles begin to cramp.

MUSCLE ACTION

The sodium-based action potential is also used to make muscles contract, by making long proteins pull on each other.

MAGNESIUM

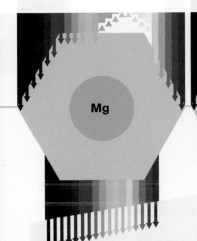

Atomic weight: 24.3050
Colour: silver–white
Phase: solid
Melting point: 650°C (1,202°F)
Boiling point: 1,090°C (1,994°F)
Crystal structure: hexagonal

Category: alkaline earth metal
Atomic number: 12

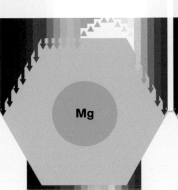

H_2O

Rubisco

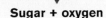

Mg

CO_2

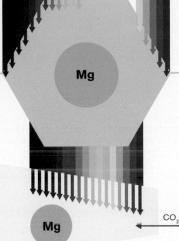

LIFE-GIVER

Magnesium is a crucial factor in the process of photosynthesis, which is the system used by plants to make sugar using the energy in sunlight. The metal is the central atom in chlorophyll, which is used by plants to collect energy from the Sun. The chlorophyll absorbs red and blue light, but reflects green light – this is why leaves always look green. The trapped energy is passed to rubisco, another magnesium chemical. This enzyme uses the energy to combine water and carbon dioxide to make glucose, the sugar that fuels all life on Earth. Oxygen is the only waste product.

Sugar + oxygen

LIFE-IMPROVER

Milk of magnesia contains magnesium hydroxide powder for settling indigestion. Talc, a soft powder used to dry the skin, is magnesium silicate.

LIGHT IN WEIGHT

Elektron is a 90-per-cent-magnesium lightweight alloy that is used in racing cars, spacecraft and airships.

LIGHT-GIVER

The bright white light of a sparkler is produced by burning magnesium powder.

ALUMINIUM

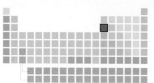

Al

13

Atomic weight: 26.981
Colour: silvery grey
Phase: solid
Melting point: 660°C (1,220°F)
Boiling point: 2,513°C (4,566°F)
Crystal structure: face-centred cubic

Category: post-transition metal
Atomic number: 13

Refinement

75% Recyling **5%**

97%

Aluminium is the most common metal in the Earth's rocks. However, it is too reactive to purify using chemical reactions, as is done with iron and copper. Instead, pure aluminium is made using a powerful electrical process. Purifying a tonne of aluminium from ore uses the same amount of energy as used by six average homes in a year. However, aluminium can be almost completely recycled – only three per cent is lost – and the process uses five per cent of the energy needed to purify the metal. Seventy-five per cent of the world's aluminium items are made from recycled metal.

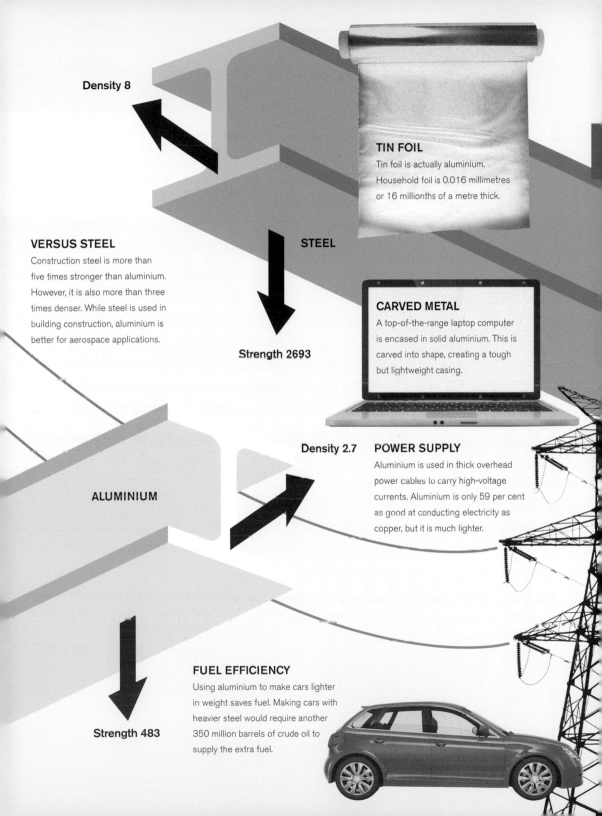

Density 8

TIN FOIL

Tin foil is actually aluminium. Household foil is 0.016 millimetres or 16 millionths of a metre thick.

STEEL

VERSUS STEEL

Construction steel is more than five times stronger than aluminium. However, it is also more than three times denser. While steel is used in building construction, aluminium is better for aerospace applications.

CARVED METAL

A top-of-the-range laptop computer is encased in solid aluminium. This is carved into shape, creating a tough but lightweight casing.

Strength 2693

Density 2.7

POWER SUPPLY

Aluminium is used in thick overhead power cables to carry high-voltage currents. Aluminium is only 59 per cent as good at conducting electricity as copper, but it is much lighter.

ALUMINIUM

FUEL EFFICIENCY

Using aluminium to make cars lighter in weight saves fuel. Making cars with heavier steel would require another 350 million barrels of crude oil to supply the extra fuel.

Strength 483

SILICON

Si 14

It is hard to overestimate the impact of silicon on human civilization over the millennia. As an ingredient in clay, it has been used to make bricks and pottery. It is an active component of the cements used to make concrete, and in the 20th century, silicon's properties as a semiconductor have revolutionized technology.

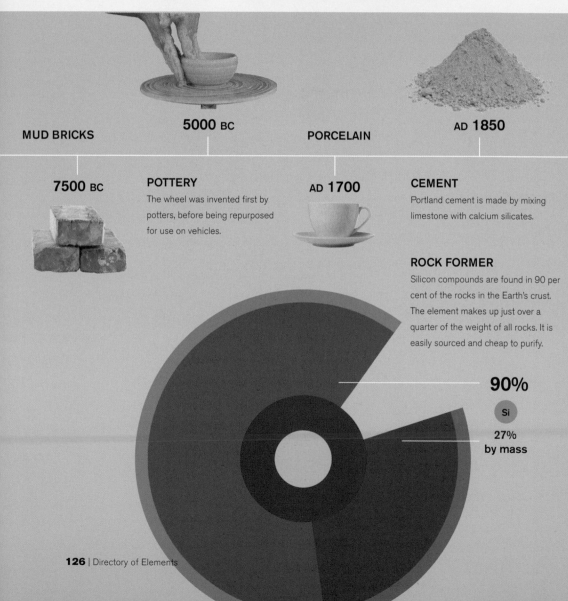

MUD BRICKS

5000 BC

PORCELAIN

AD **1850**

7500 BC

POTTERY

The wheel was invented first by potters, before being repurposed for use on vehicles.

AD **1700**

CEMENT

Portland cement is made by mixing limestone with calcium silicates.

ROCK FORMER

Silicon compounds are found in 90 per cent of the rocks in the Earth's crust. The element makes up just over a quarter of the weight of all rocks. It is easily sourced and cheap to purify.

90%

Si

27%
by mass

Atomic weight: 28.0855
Colour: metallic and bluish
Phase: solid
Melting point: 1,414°C (2,577°F)
Boiling point: 3,265°C (5,909°F)
Crystal structure: diamond cubic

Category: metalloid
Atomic number: 14

SILICONES

A polymer made from silicate molecules
is used as sealant, lubricant and
heatproof rubber.

AD **1940**

SILICON CHIPS

Pure slices, or chips, of
silicon are used to make
circuits of transistors, tiny
electronic switches that
form the basis of computer
processors.

AD **2000**

AD **1901**

MICROMECHANICALS

Tiny components, measuring a few
millionths of a metre, are created from
single crystals of silicon for use in
nanotechnology. These are machines
small enough to be placed inside
the body.

AD **1958**

SOLAR PANELS

Silicon-based solar arrays
are used to provide power
to spacecraft during
extended stays in orbit.

PHOSPHORUS

Atomic weight: 30.973762
Colour: white, black, red or violet
Phase: solid
Melting point: white: 44°C (111°F); black: 610°C (1,130°F)
Sublimation point: red: 416–590°C (781–1,094°F); violet: 620°C (1,148°F)

Boiling point: white: 281°C (538°F)
Crystal structure: white: cubic or triclinic; violet: monoclinic; black: orthorhombic; red: amorphous
Category: non-metal
Atomic number: 15

This reactive solid is the first element to have a discoverer with a known name. In 1669, the German alchemist Hennig Brand isolated pure phosphorus as a glowing white solid. As an alchemist, he had been trying to make gold from a very inexpensive substance: urine.

MAKING PHOSPHORUS

Brand followed a laborious method to convert urine to gold, but ended up producing phosphorus, which he nevertheless believed was a magical substance.

6,825 litres

HARD INSIDE

Bones and teeth are hardened with calcium phosphate crystals, which form around living cells. Urine contains a small amount of phosphorus compounds, as these cells are renewed and replaced. Brand believed that urine was yellow because it contained gold, and collected 6,825 litres of it from a company of soldiers stationed nearby.

Urine is left in the sun for several weeks until it smells rancid.

Urine is boiled until red oil appears on the surface.

Oil is left to cool until a black and white solid separate out.

Oil and black solid are heated.

16 hrs

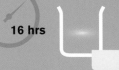

A glowing slurry is formed. Brand names it for Phosphorus, the morning star.

Substances that glow when light shines on them are called phosphors. However, they do not contain phosphorus, which glows due to a reaction with air.

SULPHUR

Atomic weight: 32.066
Colour: bright yellow
Phase: solid
Melting point: 115°C (239°F)
Boiling point: 445°C (833°F)
Crystal structure: orthorhombic

Category: non-metal
Atomic number: 16

16

S

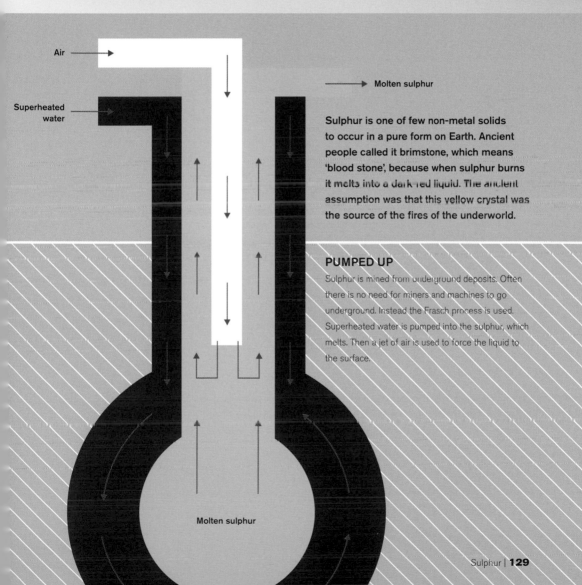

Air

Superheated water

→ Molten sulphur

Molten sulphur

Sulphur is one of few non-metal solids to occur in a pure form on Earth. Ancient people called it brimstone, which means 'blood stone', because when sulphur burns it melts into a dark-red liquid. The ancient assumption was that this yellow crystal was the source of the fires of the underworld.

PUMPED UP

Sulphur is mined from underground deposits. Often there is no need for miners and machines to go underground. Instead the Frasch process is used. Superheated water is pumped into the sulphur, which melts. Then a jet of air is used to force the liquid to the surface.

CHLORINE

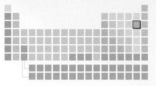

17

Cl

Atomic weight: 35.453
Colour: green–yellow
Phase: gas
Melting point: -102°C (-151°F)
Boiling point: -34°C (-29°F)
Crystal structure: n/a

Category: halogen
Atomic number: 17

Chlorine is a greenish gas that is highly reactive and corrosive.
There is little that it does not react with and kill. As a result, chlorine
is a key component of cleaning and hygiene products tasked with
killing germs. Pure chlorine is made by the electrolysis of sodium
chloride, or common salt, where a powerful electric current is used
to rip the two elements apart into their pure forms.

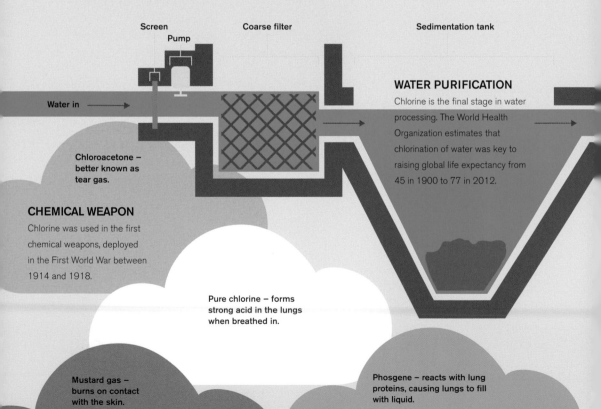

Screen

Pump

Coarse filter

Sedimentation tank

Water in

Chloroacetone –
better known as
tear gas.

WATER PURIFICATION

Chlorine is the final stage in water
processing. The World Health
Organization estimates that
chlorination of water was key to
raising global life expectancy from
45 in 1900 to 77 in 2012.

CHEMICAL WEAPON

Chlorine was used in the first
chemical weapons, deployed
in the First World War between
1914 and 1918.

Pure chlorine – forms
strong acid in the lungs
when breathed in.

Mustard gas –
burns on contact
with the skin.

Phosgene – reacts with lung
proteins, causing lungs to fill
with liquid.

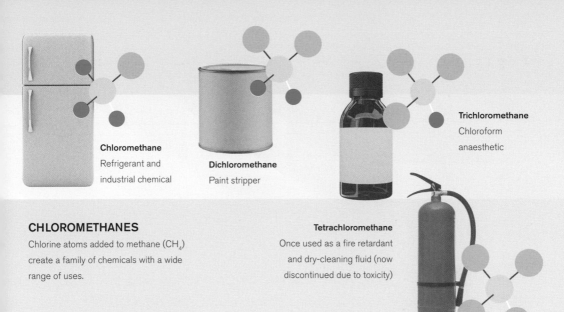

Chloromethane

Refrigerant and industrial chemical

Dichloromethane

Paint stripper

Trichloromethane

Chloroform

anaesthetic

CHLOROMETHANES

Chlorine atoms added to methane (CH_4) create a family of chemicals with a wide range of uses.

Tetrachloromethane

Once used as a fire retardant and dry-cleaning fluid (now discontinued due to toxicity)

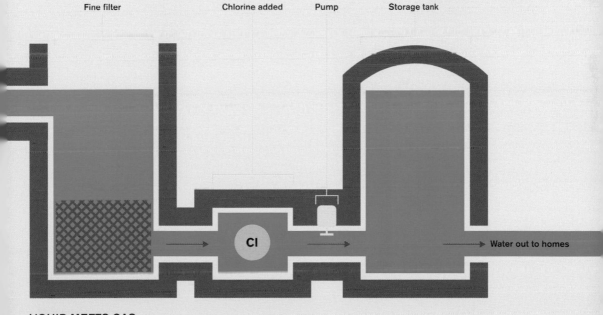

Fine filter   Chlorine added   Pump   Storage tank

Cl

Water out to homes

LIQUID MEETS GAS

Chlorination is the last phase of water processing. First a series of filters and sedimentation tanks are used to remove solid materials. Then calcium hypochlorite is added. This slowly releases tiny amounts of chlorine gas, which attack germs and give the clean water its distinctive smell.

ARGON

18	
Ar	

Atomic weight: 39.948
Colour: colourless
Phase: gas
Melting point: -189°C (-308°F)
Boiling point: -186°C (-303°F)
Crystal structure: n/a

Category: noble gas
Atomic number: 18

Argon makes up a fraction under one per cent of the air. Pneumatic chemists who studied air in the 18th and 19th centuries were consistently puzzled by a tiny portion of gas that did nothing at all. In 1894, the substance was found to be a noble gas, which was named argon, meaning 'idle'. Argon's inert nature gives it several uses. For example, it fills the spaces between the glass in double-glazed windows, cutting down the flow of heat through the window. Ancient artefacts, such as old documents, have an argon atmosphere inside their display cases. This prevents the paper being attacked by damp air, moulds and germs.

GAS SHIELD

Welding guns send a jet of argon around the hot flame, to shield it from the air and stop oxygen reacting with the items being welded.

HEAT-SEEKING MISSILE

The heat-sensitive equipment is kept cool using liquid argon.

CULLING POULTRY

Flocks of chickens that are infected with disease are killed quickly *en masse* by suffocating them in argon.

EXTINGUISHER

Argon is used as a fire suppressant in critical data centres. Other extinguishers would damage the sensitive computer equipment.

POTASSIUM

Atomic weight: 39.0983
Colour: silver–grey
Phase: solid
Melting point: 63°C (146°F)
Boiling point: 759°C (1,398°F)
Crystal structure: body-centred cubic

Category: alkali metal
Atomic number: 19

Potassium is a reactive metal found in many rocks. Like its near neighbour sodium, potassium ions play a small but crucial part in a healthy body. In fact it often works in a double-act with sodium. Potassium is constantly being lost from the body and must be replaced to ensure good health.

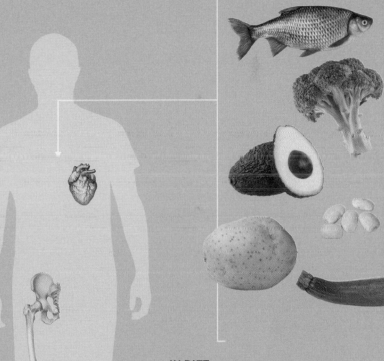

BODY SYSTEMS

Potassium is used to make the electric impulses in nerves and muscles. It is important in controlling how hard the heart pumps to create the correct blood pressure. Potassium is also involved in adding calcium to bones, and preventing it from being lost from the body. Finally, potassium ions in the blood help to control its pH so that metabolites move in and out of cells in the correct way.

IN DIET

Having a diet rich in potassium is a good way to prevent heart and nerve function problems. A lack of potassium leads to lethargy and confusion. The element is found in mushrooms, bananas, green vegetables, beans, yoghurt and fish.

CALCIUM

20

Ca

Atomic weight: 40.078
Colour: silver–grey
Phase: solid
Melting point: 842°C (1,548°F)
Boiling point: 1,484°C (2,703°F)
Crystal structure: face-centred cubic

Category: alkaline earth metal
Atomic number: 20

Stalactites

H_2O

Carbonation

CO_2

CaCO$_2$

Limestone
(calcium
carbonate)

Burning

CO_2

THE LIME CYCLE

Limestone, a natural calcium carbonate, is
a significant raw ingredient in the chemical
industry. The lime cycle is the process
for making cement, mortar and concrete.
The cycle involves a set of chemical
changes. As mortar, the material can be
used in construction by being poured as
concrete or used as a cement. Then the
mortar 'cures' and turns back into calcium
carbonate, a hard and durable solid.

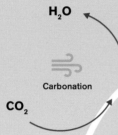

Mortar

CaO

Quicklime
(calcium oxide)

Mixing

**Sand and
water**

Ca(OH)$_2$

Slaked lime
(calcium
hydroxide)

Slaking

H_2O

MAKING DEPOSITS

Stalactites, stalagmites and other speleothems,
or cave structures, are formed when calcium
compounds dissolved in water come out of solution,
forming a solid deposit. These structures grow
slowly – about ten centimetres every 1,000 years.

Stalagmites

SCANDIUM

Atomic weight: 44.955912
Colour: silver–white
Phase: solid
Melting point: 1,541°C (2,806°F)
Boiling point: 2,836°C (5,136°F)
Crystal structure: hexagonal

Category: transition metal
Atomic number: 21

Scandium was unknown when the first periodic table was put together in 1869. However, Dmitri Mendeleev left a space for it, sure that a lightweight metal would be found at some point. Ten years later, Lars Frederik Nilson isolated a tiny sample of scandium oxide, a white powder. Nilson could prove this contained a novel element but it took until 1937 for pure scandium to be refined. Scandium has no ores – instead it appears in tiny amounts in the ores of many other metals. As a result, only about ten tonnes are mined each year.

FAST JET ALLOY

Russian MiG fighters are built using an aluminium–scandium alloy. Only a tiny amount of scandium is added but that greatly improves the alloy's strength.

REGIONAL NAME

Nilson, scandium's discoverer, was a Swede, and so he named the new element after Scandinavia. However, the main sources of the metal are in the Kola Peninsula of Russia, and in Ukraine and China.

LASER GUN

Scandium is a component in high-powered lasers developed for space warfare and as a new air-to-air weapon.

CAVITY CLEANER

Scandium lasers are also used to clean cavities. They burn away the rotted sections of tooth so it is ready to be filled.

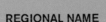

TITANIUM

22 **Ti**	

Atomic weight: 47.867
Colour: silver
Phase: solid
Melting point: 1,668°C (3,034°F)
Boiling point: 3,287°C (5,949°F)
Crystal structure: hexagonal

Category: transition metal
Atomic number: 22

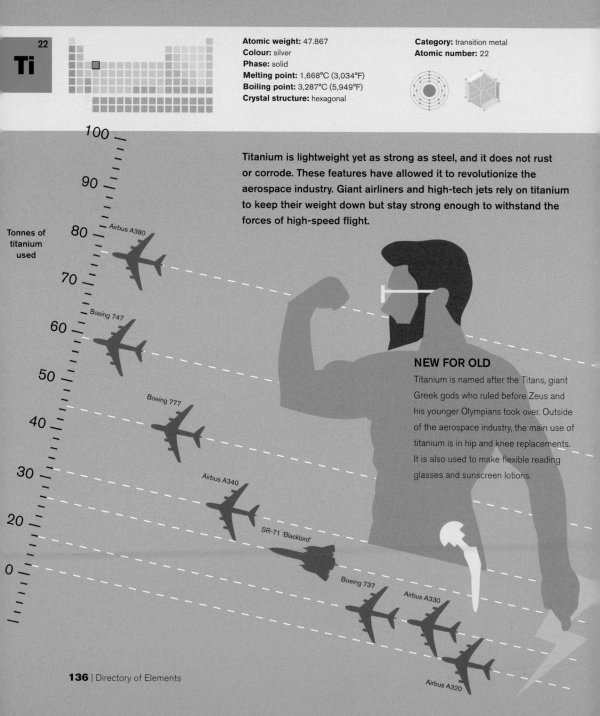

Titanium is lightweight yet as strong as steel, and it does not rust or corrode. These features have allowed it to revolutionize the aerospace industry. Giant airliners and high-tech jets rely on titanium to keep their weight down but stay strong enough to withstand the forces of high-speed flight.

Tonnes of titanium used

100
90
80 — Airbus A380
70
60 — Boeing 747
50
40 — Boeing 777
30 — Airbus A340
20 — SR-71 'Blackbird'
0 — Boeing 737
Airbus A330
Airbus A320

NEW FOR OLD

Titanium is named after the Titans, giant Greek gods who ruled before Zeus and his younger Olympians took over. Outside of the aerospace industry, the main use of titanium is in hip and knee replacements. It is also used to make flexible reading glasses and sunscreen lotions.

VANADIUM

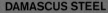

Atomic weight: 50.9415
Colour: silvery grey
Phase: solid
Melting point: 1,910°C (3,470°F)
Boiling point: 3,407°C (6,165°F)
Crystal structure: body-centred cubic

Category: transition metal
Atomic number: 23

23

V

Vanadium is produced by just three countries: Russia, China and South Africa. The metal has some very important, although niche, applications in the modern chemicals industry, and the same was true in the past – and may be true for the future.

DAMASCUS STEEL

When the Crusaders took up arms against the Saracens 1,000 years ago, they found their huge, heavy broadswords were no match for the curved sabers of the defenders. The swords and armour were made from Damascus steel. Small amounts of vanadium made this alloy very hard so it stayed sharp. In later Crusades, European soldiers used similar weapons.

V

H₂O

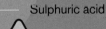

Sulphuric acid

CONTACT PROCESS

The production of sulphuric acid is a major activity in the chemical industry. The acid is made by roasting sulphur, oxygen and water. A vanadium oxide catalyst makes the reaction easier to achieve.

FUSION REACTOR

Vanadium is being used in experimental toroidal, or doughnut-shaped, fusion reactors. The metal is used because it does not expand or warp much, even when very hot. Fusion reactors recreate the nuclear energy source that powers the Sun, and it is hoped that it will be a viable power source in the coming decades.

Vanadium | **137**

CHROMIUM

24
Cr

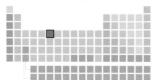

Atomic weight: 51.9961
Colour: silvery grey
Phase: solid
Melting point: 1,907°C (3,465°F)
Boiling point: 2,671°C (4,840°F)
Crystal structure: body-centred cubic

Category: transition metal
Atomic number: 24

Chromium is the shiny metal used to cover steel and other metals to prevent unsightly and damaging rust. The metal, often shortened to 'chrome', is applied using a process called electroplating.

CLEANING

The metal item to be plated with chrome is washed, polished and brushed in preparation.

ELECTRIC CURRENT

Electrical energy is used to push chromium atoms on to the item, forming a thin coating a few atoms thick.

FROM SOLUTION

The item is dipped into a bath containing a dissolved chromium compound. The current running through the item makes it negatively charged. Positively charged chromium ions are attracted to it, and the electric current gives them electrons, which converts them into metal atoms. The atoms stick to the item, forming the layer of chrome.

WASHED AND READY

The electroplated item is then washed down. The chrome layer is very hard and resists being scratched.

MANGANESE

25

Mn

Atomic weight: 54.938049
Colour: silvery grey
Phase: solid
Melting point: 1,246°C (2,275°F)
Boiling point: 2,061°C (3,742°F)
Crystal structure: body-centred cubic

Category: transition metal
Atomic number: 25

14,000,000 tonnes a year

Manganese is the fourth-most traded metal. Its main use is for producing hard steels. It is seldom produced in a pure form, but mostly as an alloy with iron or silicon. These are used as raw ingredients for making steel.

Ferromanganese 38%

30% Silicomanganese

Other alloys 8%

Slag 13%

0% Pure manganese

2%

BATTERY TECH

Manganese oxides are used in three kinds of battery: standard alkaline batteries, single-use lithium batteries (used in watches), and rechargeable lithium-ion batteries (used in mobile phones and electric cars).

IRON

Iron ore, carbon and limestone

26	
Fe	

Atomic weight: 55.845
Colour: silvery grey
Phase: solid
Melting point: 1,538°C (2,800°F)
Boiling point: 2,861°C (5,182°F)
Crystal structure: body-centred cubic

Category: transition metal
Atomic number: 26

Iron is produced by smelting. This is a reaction where the iron ore, an oxide, reacts with carbon. The carbon is oxidized into carbon dioxide, while the ore is 'reduced' into pure metal. The actual smelting reaction involves several steps that occur at different temperatures within the smelter.

CO, CO$_2$, N$_2$

23m	230°C	$3Fe_2O_3 + CO \longrightarrow 2Fe_3O_4 + CO_2$
20m	410°C	$Fe_3O_4 + CO \longrightarrow 3FeO + CO_2$
17m	525°C	$FeO + CO \longrightarrow Fe + CO_2$
14m	865°C	$C + CO_2 \longrightarrow 2CO$
11m	945°C	$CaCO_3 \longrightarrow CaO + CO_2; C + CO_2 \longrightarrow 2CO$
8m	1,125°C	$CaO + SiO_2 \longrightarrow CaSiO_3; C + CO_2 \longrightarrow 2CO$
5m	1,300°C	$C + O_2 \longrightarrow CO_2$

Air →   ← Air

Slag ←

Iron →

PIG IRON

Smelted iron contains a large amount of carbon impurities, which makes it brittle. Completely pure iron is weak and bends. However, steel, an alloy with a small and precise amount of carbon, is much stronger.

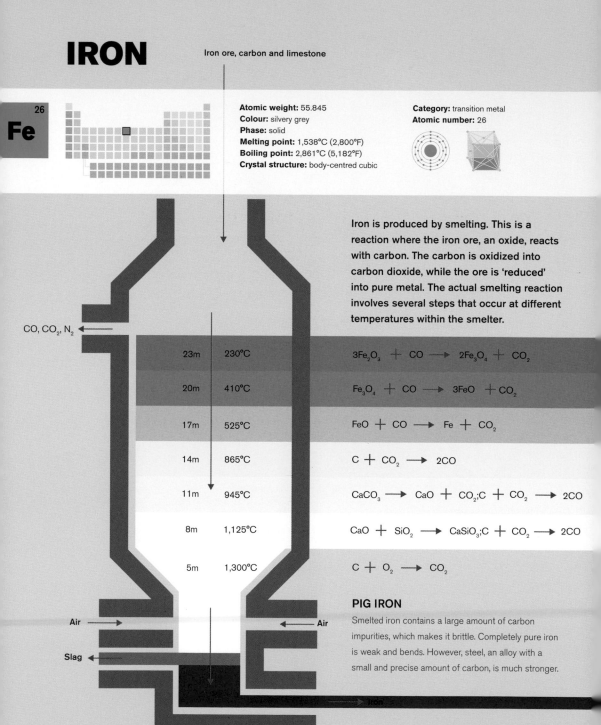

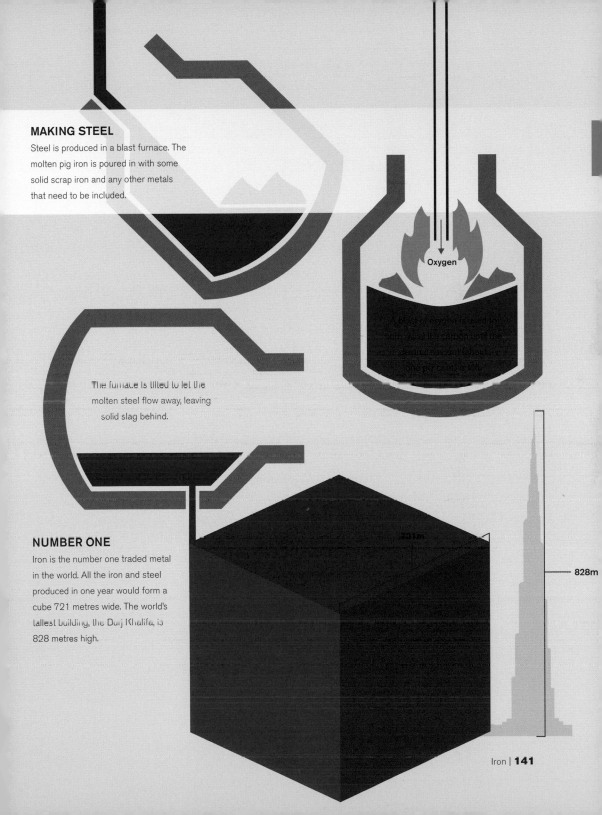

MAKING STEEL

Steel is produced in a blast furnace. The molten pig iron is poured in with some solid scrap iron and any other metals that need to be included.

Oxygen

A blast of oxygen is used to burn away the carbon until the desired amount (about one per cent) is left.

The furnace is tilted to let the molten steel flow away, leaving solid slag behind.

NUMBER ONE

Iron is the number one traded metal in the world. All the iron and steel produced in one year would form a cube 721 metres wide. The world's tallest building, the Burj Khalifa, is 828 metres high.

721m

828m

COBALT

COBALT GREEN

Doping zinc oxide or similar white chemicals gives them a pale green colour. Cobalt impurities give some gems their green colour.

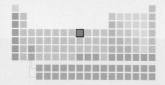

Atomic weight: 58.9332
Colour: metallic grey
Phase: solid
Melting point: 1,495°C (2,723°F)
Boiling point: 2,927°C (5,301°F)
Crystal structure: hexagonal

27
Co

Category: transition metal
Atomic number: 27

Cobalt ores were feared by ancient miners, who named them after devlish goblins, or kobolds. The problem was that the ores resemble silver-bearing minerals but were, in fact, cobalt arsenide, which gave out toxic fumes when smelted. However, cobalt did find a use in a variety of pigments that bear its name.

COBALT BLUE

Made from colbalt aluminate, this pigment was traditionally used in Chinese porcelains.

CERULEAN BLUE

Made from cobalt stannate, a compound with tin and oxygen, this blue was a favourite with 19th-century impressionist painters – along with cobalt blue.

COBALT VIOLET

Invented in 1859, this was the first stable violet pigment. Based in cobalt phosphate, it was developed by the 19th-century colourist, Louis Alphonse Salvètat.

AUREOLIN

Also known as cobalt yellow, this pigment contains potassium cobaltinitrite. It was used in small amounts due to its great expense.

NICKEL

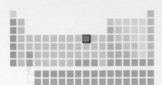

Atomic weight: 58.6934
Colour: silvery white
Phase: solid
Melting point: 1,455°C (2,651°F)
Boiling point: 2,912°C (5,274°F)
Crystal structure: face-centred cubic

Category: transition metal
Atomic number: 28

28
Ni

Nickel is used chiefly as an ingredient of stainless steels and other high-tech alloys. As plating, nickel is a cheap alternative to silver. The metal is also used in batteries, electronics and, of course, in coinage.

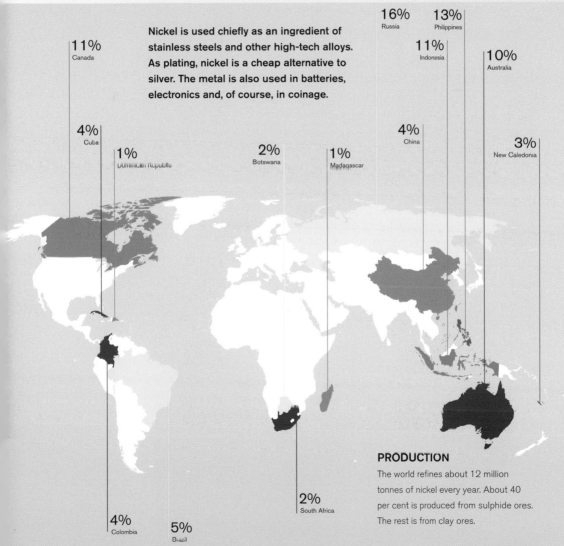

16%
Russia

13%
Philippines

11%
Indonesia

10%
Australia

11%
Canada

4%
China

3%
New Caledonia

4%
Cuba

1%
Dominican Republic

2%
Botswana

1%
Madagascar

4%
Colombia

5%
Brazil

2%
South Africa

PRODUCTION

The world refines about 12 million tonnes of nickel every year. About 40 per cent is produced from sulphide ores. The rest is from clay ores.

COPPER

8000 BC
Beads in jewellery

5000 BC
Timna Mine

3000 BC
Bronze Age

29
Cu

Copper was the first mass-produced metal, with copper mining dating back 7,000 years. Even before this, naturally pure copper was being worked into jewels. After centuries of innovation, copper was given a new role with the rise of electrical technology in the 19th century.

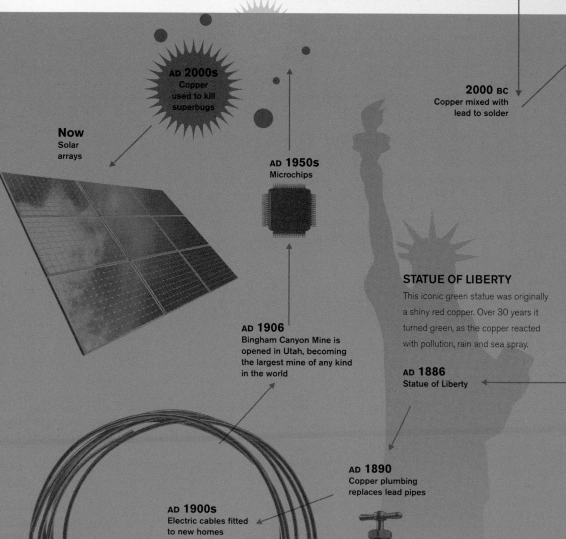

AD 2000s
Copper used to kill superbugs

2000 BC
Copper mixed with lead to solder

Now
Solar arrays

AD 1950s
Microchips

STATUE OF LIBERTY

This iconic green statue was originally a shiny red copper. Over 30 years it turned green, as the copper reacted with pollution, rain and sea spray.

AD 1906
Bingham Canyon Mine is opened in Utah, becoming the largest mine of any kind in the world

AD 1886
Statue of Liberty

AD 1890
Copper plumbing replaces lead pipes

AD 1900s
Electric cables fitted to new homes

1000 BC
Coins

AD 900
Falun Mine

Atomic weight: 63.546
Colour: red–orange
Phase: solid
Melting point: 1,085°C (1,985°F)
Boiling point: 2,562°C (4,644°F)
Crystal structure: face-centred cubic

Category: transition metal
Atomic number: 29

COPPER MOUNTAIN

Falun Mine, Sweden, was the primary source of Europe's copper in medieval times. The mine was worked from the 10th century until 1992.

480 BC
Bronze rams fitted to Greek ships sink the Persian fleet at the Battle of Salamis

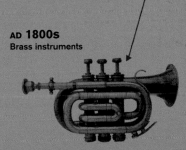

AD 1500s
Copper bottoms are added to ships to protect the wood from shipworms on long voyages

AD 1839
A daguerrotype camera captures an image on a sheet of copper

AD 1730
Brass mills mix copper with zinc to make a tough, golden alloy

AD 1800s
Brass instruments

AD 1830
Electromagnets create a controllable magnetic field, with copper wire coiled around an iron core

ZINC

30		
Zn		

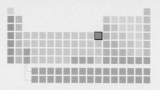

Atomic weight: 65.409
Colour: bluish-white
Phase: solid
Melting point: 420°C (788°F)
Boiling point: 907°C (1,665°F)
Crystal structure: hexagonal

Category: transition metal
Atomic number: 30

Pure zinc was not identified until 1746, but zinc minerals have been in use for many centuries. For example, calamine lotion, an ancient remedy for soothing itchy chicken pox, contains zinc oxide. A similar zinc chemical is added to shampoos to tackle dandruff.

SACRIFICIAL PROTECTOR

Zinc is more reactive than other transition metals, like iron. For that reason, zinc can be used to protect steel in a sacrificial system.

Galvanized steel is coated in a layer of zinc. When the metal is scratched, the steel is exposed to air and water and may rust. However, the zinc reacts before the steel metals, sealing the scratch with a compound and keeping the steel safe from corrosion.

Zinc

Steel

PALE PROTECTOR

Zinc oxide is a good reflector of light – that's why it's a bright white. The oxide is used in sunscreens and in space suits. In both cases it is there to reflect away damaging radiation.

GALLIUM

Atomic weight: 69.723
Colour: silvery blue
Phase: solid
Melting point: 30°C (86°F)
Boiling point: 2,204°C (3,999°F)
Crystal structure: orthorhombic

Category: post-transition metal
Atomic number: 31

31
Ga

Gallium was the first element to be named after a country – perhaps. Paul Émile Lecoq named it for Gallia, the Latin for France, in 1875. However, gallia is derived from *gallus*, the Latin for 'rooster' or *le coq* in French. Commentators suggested that Lecoq was actually naming the metal after himself.

SOFT TOUCH

Gallium is a soft solid in standard conditions, but it will melt when held in the hand. Body temperature is above its melting point.

SAFER LIQUID

The first accurate thermometers used mercury. However, handling mercury is always a risk, so the medical thermometers used to take accurate body temperatures use galinstan. This is a mixture of gallium, indium and tin, which is liquid above −19°C.

GERMANIUM

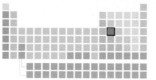

32

Ge

Atomic weight: 72.631
Colour: grey–white
Phase: solid
Melting point: 938°C (1,720°F)
Boiling point: 2,833°C (5,131°F)
Crystal structure: diamond cubic

Category: metalloid
Atomic number: 32

In the original 1869 periodic table, there was no known element 32. Mendeleev predicted that it existed, however, and set out the properties of this unknown element, which he called ekasilicon. In 1886, the element was discovered and named germanium. Its properties tallied very closely with Mendeleev's predictions, and the scientific world woke up to the power of the periodic table.

	Eka	Ge
Atomic mass	72.64	72.63
Density (g/cm³)	5.5	5.35
Melting point (°C)	high	938
Colour	grey	grey
Oxide type	refractory dioxide	refractory dioxide
Oxide density (g/cm³)	4.7	4.7
Oxide activity	feebly basic	feebly basic
Chloride boiling point (°C)	under 100	86 ($GeCl_4$)
Chloride density (g/cm³)	1.9	1.9

SOUND AND VISION

Germanium is a semiconductor, and that property sees it being used in a range of high-tech applications. The laser that burns data onto a writable DVD uses germanium. Night-vision goggles use it to convert infrared into a visible image, and optical fibres are doped with germanium oxide to help them keep laser signals reflecting inside.

ARSENIC

Atomic weight: 74.92160
Colour: grey
Phase: solid
Melting point: n/a
Sublimation point: 614°C (1,137°F)
Crystal structure: trigonal

Category: metalloid
Atomic number: 33

Arsenic minerals are often impressive, with a metallic sheen or bright colours. As a result they have traditionally been used as pigments, especially in golden paints. However, pure arsenic and its oxides are toxic, and have a long history of being a slow but steady method for poisoning people. The minerals have a garlic smell, and therefore may at least improve the flavour of a final meal!

NAPOLEON

Large amounts of arsenic were found in the hair of Napoleon Bonaparte after his death in 1821. Was he poisoned, or was his plush, green wallpaper emitting a deadly arsenic vapour?

DEADLY TREATS

Sweets sold on a market stall in Bradford, England, in 1858 were contaminated with arsenic; 200 people fell ill and 21 died.

MARY ANN COTTON

From 1852 to 1873, this serial killer from Sunderland, England, used arsenic to kill four husbands, 13 of her children and two lovers.

TOASTED TO DEATH

The Borgias family, who ruled much of southern Europe in the 15th and 16th centuries, frequently got rid of enemies by serving wine poisoned with arsenic.

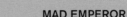

LOST EMPEROR

In 1908, the Guangxu emperor of China died from a stomach ache – and his face turned blue. This all suggests he was poisoned with arsenic, probably by corrupt civil servants (who were generally eunuchs). His nephew, Puyi, became China's last emperor.

MAD EMPEROR

In AD 55, Nero ordered the death of his little brother Brittanicus using arsenic. That cleared the way for Nero to become emperor.

SELENIUM

Selenium was named for Selene, the Greek goddess of the moon.

34
Se

Atomic weight: 78.96
Colour: metallic grey
Phase: solid
Melting point: 221°C (430°F)
Boiling point: 685°C (1,265°F)
Crystal structure: hexagonal

Category: non-metal
Atomic number: 34

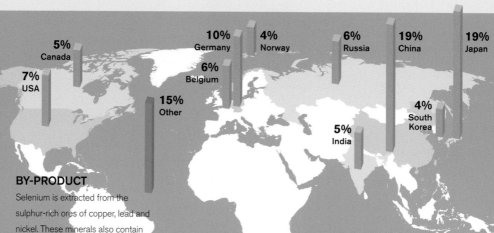

5% Canada

7% USA

10% Germany

6% Belgium

4% Norway

15% Other

6% Russia

19% China

19% Japan

4% South Korea

5% India

BY-PRODUCT

Selenium is extracted from the sulphur-rich ores of copper, lead and nickel. These minerals also contain a small amount of this element.

2,670 TONNES

Metallurgical
35%

Feed
10%

Batteries
10%

Glass
25%

Electrical
10%

Other
10%

Pure selenium often has a metallic sheen, but the element is a non-metal. It is used in many industries, including as a food supplement. Animals need small amounts in their diets for good health.

BROMINE

Bromine is the only non-metal element that is a liquid. These charts compare it with water, a more familiar liquid.

Atomic weight: 79.904
Colour: deep red
Phase: liquid
Melting point: -7°C (19°F)
Boiling point: 59°C (138°F)
Crystal structure: orthorhombic

Category: halogen
Atomic number: 35

35

Br

ORANGE GAS

Bromine boils into a choking vapour at a lower temperature than water. The burning smell of the gas earned the element its name, which means 'stench'.

H₂O

Br

| 100°C | 59°C |

BROWN LIQUID

Bromine and water mix – they dissolve in each other. However, bromine is much more dense, and so sinks to the bottom of the container.

H₂O

Br

Viscosity

Bromine and water have a similar viscosity, so they flow and splash in the same way.

| 0.9 | 0.95 |

Density

| 1 | 3.11 |

YELLOW SOLID

Bromine and water freeze at similar temperatures.

H₂O

Br

| 0°C | -7°C |

10,600 —

Reprocessing
1949–present

KRYPTON

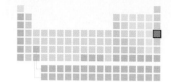

36

Kr

Atomic weight: 83.798
Colour: colourless
Phase: gas
Melting point: -157°C (-251°F)
Boiling point: -153°C (-244°F)
Crystal structure: n/a

Category: noble gas
Atomic number: 36

180 —

Weapons tests
1945–present

160 —

Natural krypton is very rare in the air.
However, the krypton-85 isotope is
produced by nuclear fission, and is
used as a measure of nuclear activities.
Sampling the levels of Kr-85 reveal
weapons testing, nuclear accidents and
the impact of nuclear-waste processing.

 46%

RELEASE LIGHTNING

The emission of Kr-85 above nuclear
facilities changes the conductivity of
the air, and leads to a major spike in the
number of lightning strikes in the area.

140 —

120 —

100 —

Fukushima
2011

80 —

60 —

40 —

Chernobyl
1986

Three Mile Island
1979

20 —

0 —

PBq

RUBIDIUM

Atomic weight: 85.4678
Colour: silvery white
Phase: solid
Melting point: 39°C (102°F)
Boiling point: 688°C (1,270°F)
Crystal structure: body-centred cubic

Category: alkali metal
Atomic number: 37

In 1995, rubidium was made the coldest thing in the Universe. It was chilled to 0.001K (−273.14°C), just above absolute zero, and a few degrees colder than deep space. At this temperature, the rubidium atoms gave up their individual identity and fused into a Bose–Einstein condensate. This is a state of matter where atoms disappear. It had been predicted in 1924, but the technology needed to make it took some time to catch up.

LASER COOLING

After using powerful refrigerators to chill rubidium gas, the temperature was reduced some more using lasers. Atoms absorb laser light, and if the laser is angled just right, it can slow an atom's motion – and that makes the gas cooler, atom by atom.

MAGNETIC TRAP

The gas was held in a magnetic-field 'bowl'. Warmer rubidium atoms were moving fast enough to escape from the bowl. Gradually the bowl was shrunk, leaving only the very coldest atoms at the bottom. These transformed into the condensate made of the matter in just a few hundred atoms.

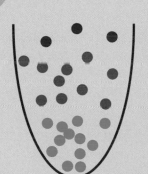

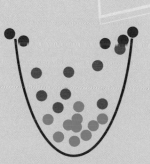

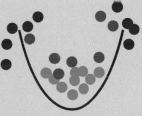

STRONTIUM

550,000

500,000

38

Sr

Atomic weight: 87.62
Colour: silver–grey
Phase: solid
Melting point: 777°C (1,431°F)
Boiling point: 1,382°C (2,520°F)
Crystal structure: face-centred cubic

Category: alkaline earth metal
Atomic number: 38

400,000

Most strontium is produced by China, Mexico, Spain and Argentina. During the late 20th century, three-quarters of it was used in TV screens. A coating of strontium stopped the cathode ray inside old-fashioned sets from emitting X-rays. Today's TVs work using LCDs and do not need strontium, and production dipped sharply in 2005. However, strontium is now being used more in drilling muds, which are heavy slurries used to stop gases bursting from oil wells.

Cathode-ray TV

350,000

300,000

OTHER USES

Strontium compounds produce the red smoke in warning flares. It strengthens magnets and is the active ingredient in toothpaste for sensitive teeth. Blue pigments often contain strontium, and the metal is involved in glassmaking, alloys and zinc refining.

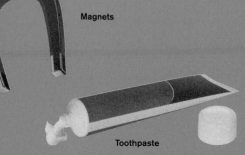

Drilling mud

250,000

Paint

200,000

150,000

Magnets

Flare

Glass

Zinc   Alloys

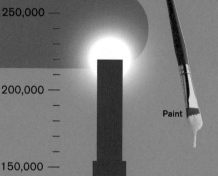

100,000

Toothpaste

0

1995   2000   2005   2010   2015

YTTRIUM

Atomic weight: 88.90585
Colour: silver–white
Phase: solid
Melting point: 1,526°C (2,779°F)
Boiling point: 3,336°C (6,037°F)
Crystal structure: hexagonal

Category: transition metal
Atomic number: 39

39

Y

Yttrium is crystallized with aluminium garnet, making YAG, the primary source of lasers. YAG lasers are used in eye surgery, tattoo removal, range finders and welding.

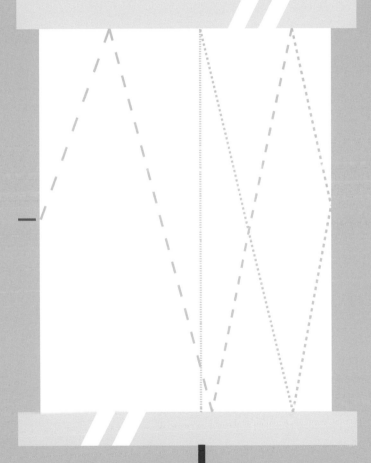

Light

AMPLIFYING LIGHT

YAG crystals are the gain medium of the laser. Light shines into the crystal and that energy excites the atoms, making them release more light with a specific wavelength, or colour. The crystal has mirrors at each end, so the light bounces around inside – and that makes the atoms produce even more light. The amplified light is then released as a pulse, or beam, through a hole in one mirror.

Laser

tonnes

RUS UKR GMB ZAF USA BRA CHN IND IDN MYS LKA AUS

0
25,000
50,000
100,000
125,000
150,000
175,000
200,000
225,000
250,000
275,000
300,000
325,000
350,000
375,000
400,000
425,000
450,000

ZIRCONIUM

40
Zr

Atomic weight: 91.224
Colour: silvery white
Phase: solid
Melting point: 1,855°C (3,371°F)
Boiling point: 4,409°C (7,968°F)
Crystal structure: hexagonal close-packed

Category: transition metal
Atomic number: 40

Zirconium's most familiar compound is cubic zirconia, which is used as a substitute for diamonds. The two crystals look very similar but there are differences.

COLOUR

Zirconia crystals are clearer than most diamonds, which have a yellow or brownish hue.

HEAT TRANSFER

Diamond allows heat to pass through it, but zirconia is an insulator. As well as gems, the crystals are used in heat-resistant ceramics.

10 — Diamond

9

8 — Cubic zirconia

HARDNESS

Diamond is the hardest natural substance known to man. Zirconia cannot compete with it, but it is still a very hard substance.

DIRTY SECRET

The giveaway feature of a diamond is that it still sparkles when dirty; a zirconia will not.

HEAVY STUFF

A zirconia is 1.7 times denser than a diamond.

OLD STUFF

Zircon, a silicate of zirconium, is the oldest material on Earth. Zircon crystals from Australia are 4.4 billion years old.

NIOBIUM

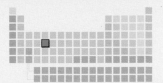

Atomic weight: 92.90638
Colour: steel grey
Phase: solid
Melting point: 2,477°C (4,491°F)
Boiling point: 4,744°C (8,571°F)
Crystal structure: body-centred cubic

Category: transition metal
Atomic number: 41

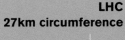

41

Nb

LHC
27km circumference

1,200 tonnes

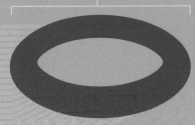

Niobium is the 34th most common element in the Earth's crust, but it is widely distributed and has no major ore. Niobium has several important uses in electronics, especially capacitors. Every smartphone contains a little of this metal. Niobium's production has doubled in recent years, but is seldom more than 50,000 tonnes a year.

SUPERCONDUCTOR

The most intensive use of niobium is to make superconducting alloy wires. These are used by particle accelerators, such as the Large Hadron Collider (LHC) at CERN, to provide power to the electromagnets that control the motion of particle beams. The ITER fusion reactor, currently being built in France, is set to be the most concentrated mass of niobium in history.

Tevatron
6.8km circumference

17 tonnes

ITER
0.09km circumference

850 tonnes

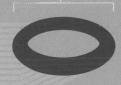

MOLYBDENUM

42		
Mo		

Atomic weight: 95.94
Colour: silvery white
Phase: solid
Melting point: 2,623°C (4,753°F)
Boiling point: 4,639°C (8,382°F)
Crystal structure: body-centred cubic

Category: transition metal
Atomic number: 42

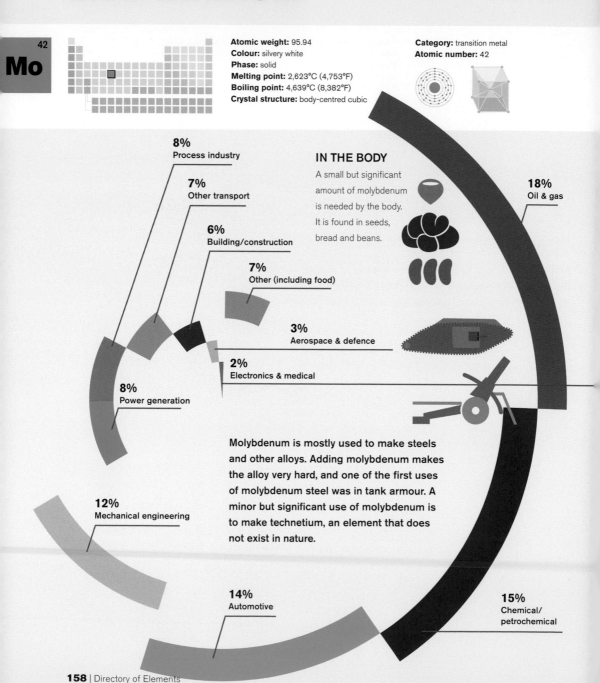

8%
Process industry

7%
Other transport

6%
Building/construction

7%
Other (including food)

IN THE BODY

A small but significant amount of molybdenum is needed by the body. It is found in seeds, bread and beans.

18%
Oil & gas

3%
Aerospace & defence

2%
Electronics & medical

8%
Power generation

Molybdenum is mostly used to make steels and other alloys. Adding molybdenum makes the alloy very hard, and one of the first uses of molybdenum steel was in tank armour. A minor but significant use of molybdenum is to make technetium, an element that does not exist in nature.

12%
Mechanical engineering

14%
Automotive

15%
Chemical/
petrochemical

TECHNETIUM

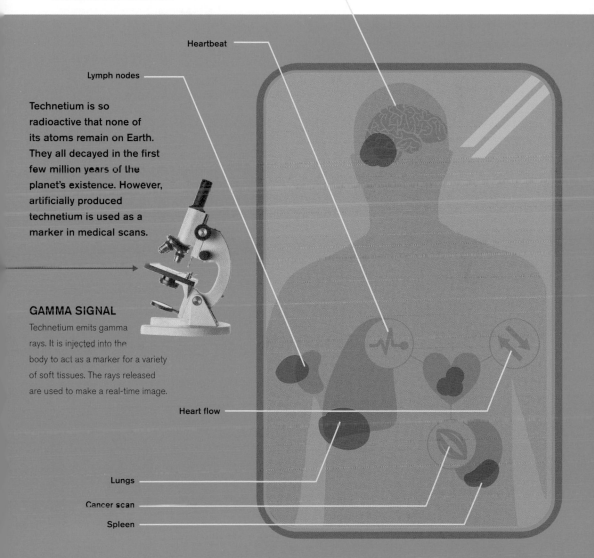

Atomic weight: 98
Colour: silvery grey
Phase: solid
Melting point: 2,157°C (3,915°F)
Boiling point: 4,265°C (7,709°F)
Crystal structure: hexagonal

Category: transition metal
Atomic number: 43

43

Tc

Brain

Heartbeat

Lymph nodes

Technetium is so radioactive that none of its atoms remain on Earth. They all decayed in the first few million years of the planet's existence. However, artificially produced technetium is used as a marker in medical scans.

GAMMA SIGNAL

Technetium emits gamma rays. It is injected into the body to act as a marker for a variety of soft tissues. The rays released are used to make a real-time image.

Heart flow

Lungs

Cancer scan

Spleen

RUTHENIUM

44

Ru

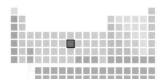

Atomic weight: 101.07
Colour: silvery white
Phase: solid
Melting point: 2,334°C (4,233°F)
Boiling point: 4,150°C (7,502°F)
Crystal structure: hexagonal

Category: transition metal
Atomic number: 44

Raw materials
(natural gas/coal)

Ruthenium is a very rare
metal. It is used in tiny
amounts to make high-tech
alloys, but its main use is
as a catalyst. One of the
reactions it facilitates is the
Fischer-Tropsch process,
which converts coal and
natural gas into liquid
hydrocarbon fuels.

RESOURCE REPURPOSE

The Fischer-Tropsch process is used
in regions that do not have access
to petroleum. It converts the carbon
compounds in coal and gas into
useful liquid hydrocarbons.

SYNGAS

The raw materials are reacted with
oxygen in a controlled way, which makes
syngas. This is a flammable mixture of
hydrogen and carbon monoxide.

Syngas

ENTER RUTHENIUM

A series of catalysts, including
ruthenium, combine the carbon
monoxide and hydrogen into long-chain
hydrocarbons, such as octanes. These
are flammable liquids that fuel vehicles
and can be used as raw materials for
drugs and other chemicals.

Fischer-Tropsch process

End product
(mostly fuel)

RHODIUM

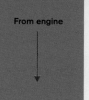

Atomic weight: 102.90550
Colour: silvery white
Phase: solid
Melting point: 1,964°C (3,567°F)
Boiling point: 3,695°C (6,683°F)
Crystal structure: face-centred cubic

Category: transition metal
Atomic number: 45

45

Rh

Rhodium is the second-rarest element in Earth's rocks – of the elements that can be mined and refined, that is. There are less than three rhodium atoms among every billion making up the crust. However, despite that, most people own a bit of it. Minute amounts of rhodium are at work in the catalytic converters in every car.

Rh

REDUCTION CATALYST

A catalytic converter turns the most toxic and polluting gases in car exhaust into less harmful ones. The rhodium is responsible for reducing nitrogen oxides. It reacts them with carbon monoxide in the exhaust to make nitrogen and carbon dioxide. A palladium catalyst does the opposite; it oxidizes unburned hydrocarbons into water and carbon dioxide.

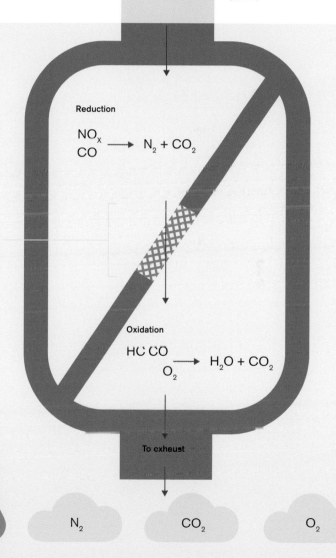

Reduction

$$NO_x + CO \longrightarrow N_2 + CO_2$$

Oxidation

$$HC\ CO + O_2 \longrightarrow H_2O + CO_2$$

To exhaust

H_2O N_2 CO_2 O_2

PALLADIUM

46

Pd

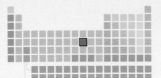

Atomic weight: 106.42
Colour: silvery white
Phase: solid
Melting point: 1,555°C (2,831°F)
Boiling point: 2,963°C (5,365°F)
Crystal structure: face-centred cubic

Category: transition metal
Atomic number: 46

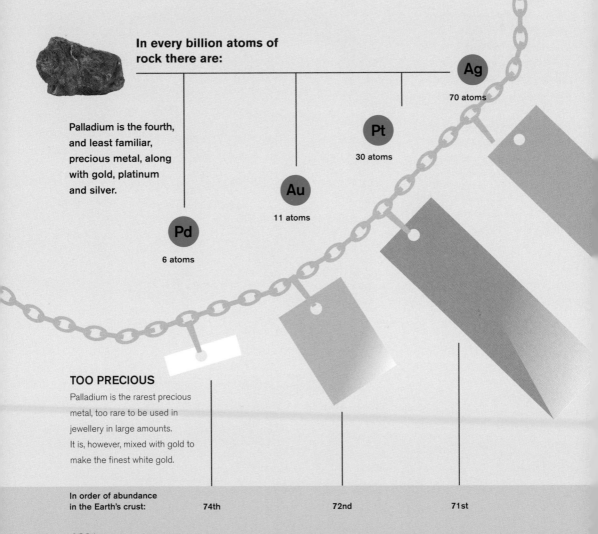

In every billion atoms of rock there are:

Palladium is the fourth, and least familiar, precious metal, along with gold, platinum and silver.

Ag
70 atoms

Pt
30 atoms

Au
11 atoms

Pd
6 atoms

TOO PRECIOUS

Palladium is the rarest precious metal, too rare to be used in jewellery in large amounts. It is, however, mixed with gold to make the finest white gold.

In order of abundance in the Earth's crust:

74th   72nd   71st

SILVER

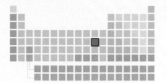

Atomic weight: 107.8682
Colour: brilliant-white metallic
Phase: solid
Melting point: 962°C (1,764°F)
Boiling point: 2,162°C (3,924°F)
Crystal structure: face-centred cubic

Category: transition metal
Atomic number: 47

47

Ag

Silver has been used for at least 5,000 years. Native, or naturally pure, silver forms in places where mineral-rich waters trickle through rocks. Silver is the most reactive of the precious metals, and a sensitivity to light meant that silver was instrumental in the development of photography.

IMAGE CAPTURE

Light hits a grain of silver bromide.

The light's energy splits a few of the silver bromides into separate silver and bromine atoms.

A developer chemical reduces the silver bromide into silver and bromine, but only in the grains that already have silver atoms present.

The unexposed grains are washed away and patches of pure silver create a negative image, where light regions appear dark.

65th

CADMIUM

Cd 48

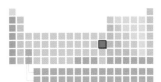

Atomic weight: 112.411
Colour: silvery blue
Phase: solid
Melting point: 321°C (610°F)
Boiling point: 767°C (1,413°F)
Crystal structure: hexagonal

Category: transition metal
Atomic number: 48

Ångström

Cadmium is a soft and highly toxic metal. A lengthy exposure leads to sensitive and painful joints, a condition called *itai-itai*, which means 'ouch-ouch' in Japanese. Before its toxicity was appreciated, cadmium sulphide was used to make yellow oil paints. They were popular with Van Gogh, Matisse and Monet. However, it is cadmium's ability to release red light that is used today. The wavelength of that red light is used to define a tiny unit called the ångström (Å).

WAVELENGTHS

The human eye shows the wavelength of light as colour. The eye can detect light between 4,000 (blue) and 7,000 Å (red). X-rays, which carry much more energy, have a wavelength of about 1 Å.

SMALL BUT USEFUL

The wavelength of light and other radiation is generally a very tiny distance. In 1868, it was suggested that we use a new unit to measure it, and the result was the ångström, which is ten billionths of a metre wide. But how do you measure such a distance? In 1907, it was decided that the red line in cadmium's emission spectrum would be exactly 6438.46963 Å. Cadmium was chosen because its line was easy to identity, and that became the benchmark from which every other wavelength is measured.

INDIUM

Atomic weight: 114.818	**Category:** post-transition metal
Colour: silvery grey	**Atomic number:** 49
Phase: solid	
Melting point: 157°C (314°F)	
Boiling point: 2,072°C (3,762°F)	
Crystal structure: tetragonal	

49

In

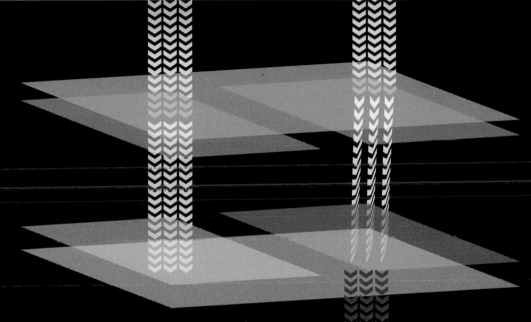

While strontium was once in high demand for old-fashioned television screens, today it is indium that is the ubiquitous constituent. Indium tin oxide is the conductor that carries a current to the pixels that create the coloured dots on LCD screens. Indium tin oxide is chosen because it can be made into layers so thin that the light shines right through.

The name indium is derived from the colour indigo, a purple dye that originally hails from India. Indium produces a clear indigo line when electrified.

TIN

50
Sn

Atomic weight: 118.710
Colour: silver–white
Phase: solid
Melting point: 232°C (449°F)
Boiling point: 2,602°C (4,716°F)
Crystal structure: tetragonal

Category: post-transition metal
Atomic number: 50

50

IS THE MAGIC NUMBER

Tin has 50 protons in its nucleus. This imparts a great stability on the nucleus, with each proton locked together into 25 pairs. As a result, tin has the most number of stable isotopes of any element. The isotopes of an element all have the same number of protons but a varying number of neutrons. All elements have a range of isotopes and most are highly radioactive and short-lived. Tin, on the other hand, has ten stable isotopes with between 62 and 74 neutrons.

Natural abundance

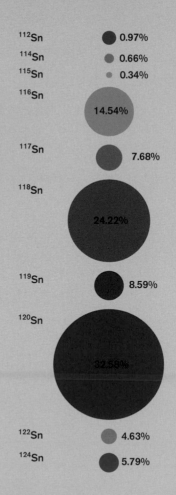

^{112}Sn   0.97%
^{114}Sn   0.66%
^{115}Sn   0.34%
^{116}Sn   14.54%
^{117}Sn   7.68%
^{118}Sn   24.22%
^{119}Sn   8.59%
^{120}Sn   32.58%
^{122}Sn   4.63%
^{124}Sn   5.79%

ANTIMONY

Tonnes

1,000,000
950,000
900,000
850,000
800,000
750,000
700,000
650,000
600,000
550,000
500,000
450,000
400,000
350,000
300,000
250,000
200,000
150,000
100,000
50,000
0

51
Sb

Atomic weight: 121.760
Colour: silver–grey
Phase: solid
Melting point: 631°C (1,168°F)
Boiling point: 1,587°C (2,889°F)
Crystal structure: trigonal

Category: metalloid
Atomic number: 51

The minerals of this silvery metalloid have a long history. As stibnite, a sulphide compound, it was used as eye make-up in ancient Egypt. Stibnite is now the major antimony ore. However, the proven global reserves are running low.

Total world reserves
1,987,000
tonnes

CHN
(47.81%)

Annual world production
180,000
tonnes

Antimony strengthens the lead in car batteries. It also clears air bubbles from ultra high-spec glass, and antimony trioxide is a fire retardant.

RUS
(17.61%)

BOL
(15.6%)

Other countries
(5.03%)

AUS
(7.05%)

ZAF
(1.36%)

TJK
(2.52%)

USA
3.02%)

11

TELLURIUM

52

Te

Atomic weight: 127.60
Colour: silver–white
Phase: solid
Melting point: 449°C (841°F)
Boiling point: 988°C (1,810°F)
Crystal structure: hexagonal

Category: metalloid
Atomic number: 52

Tellurium has an interesting link to light. It is used to produce a range of glazing effects on ceramics. The CCD (charge coupled device) that captures an image for a digital camera contains the semimetal, and cadmium telluride is used to make solar panels that are much cheaper (but also less efficient) than silicon ones.

01
010100100
10101010101
010010010
01010

FOOLISH GOLD MINERS

In the 1893 Kalgoorlie gold rush, the dark, shiny mineral called calaverite was discarded by miners, who thought it was fools' gold. They used it as ballast for new roads to the mines. In 1896, it was found that calaverite was in fact gold telluride. The miners rushed back and dug up the roads.

IODINE

Brain

Atomic weight: 126.90447
Phase: solid
Colour: black
Melting point: 114°C (237°F)
Boiling point: 184°C (364°F)
Crystal structure: orthorhombic

Category: halogen
Atomic number: 53

53

I

Iodine is an essential component of diet. In many parts of the world there is not enough natural iodine in the soil, so iodine is added to salt as a universal supplement. A deficiency in iodine in early life leads to problems with brain development. In later life it can create a goitre, a throat swelling caused by an enlarged thyroid gland.

Thyroid

■ = Risk of adverse health consequences

■ = Risk of iodine-induced hyperthyroidism

■ = Optimal iodine nutrition

□ = Mild iodine deficiency

■ = Moderate iodine deficiency

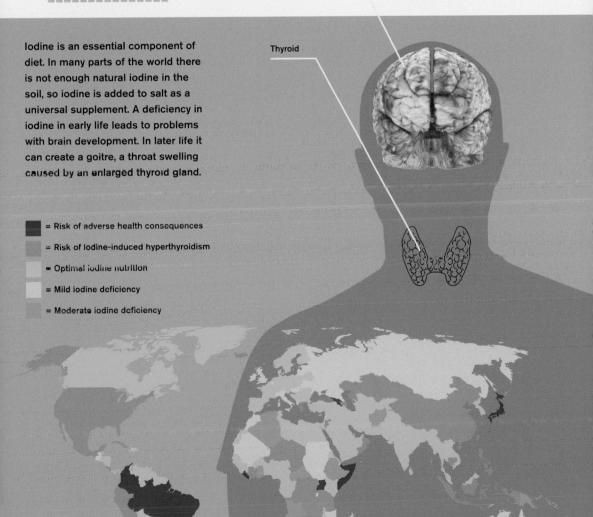

10,000

9,500

XENON

9,000

8,500

Atomic weight: 131.29
Colour: colourless
Phase: gas
Melting point: -112°C (-169°F)
Boiling point: -108°C (-163°F)
Crystal structure: n/a

Category: noble gas
Atomic number: 54

8,000

7,500

7,000

IMAX BULB

The xenon bulbs used in Imax projectors are

3,000

6,500

6,000

times brighter than regular lightbulbs. The
pressure of the gas inside is 25 times higher
than the atmosphere. Technicians must wear a
bomb-disposal suit when they change a bulb to
protect them if the bulb cracks.

5,500

5,000

4,500

4,000

3,500

3,000

Xenon is the heaviest stable noble gas. Like other
members of this group, xenon is used in gas-
discharge lamps, or 'neon lights'. Xenon produces
a 'warmer' light than regular domestic lightbulbs.
In other words, xenon bulbs produce a whiter
glow with less yellow light that looks more like
daylight. As a result, xenon bulbs are used in
camera flashes, high-end car headlights and
film projectors.

2,500

2,000

1,500

1,000

CAESIUM

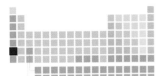

Atomic weight: 132.90545
Colour: silvery gold
Phase: solid
Melting point: 28°C (83°F)
Boiling point: 671°C (1,240°F)
Crystal structure: body-centred cubic

Category: alkali metal
Atomic number: 55

CS

1. An oven heats caesium to create a stream of ions.

Atomic clocks use caesium to keep time, and can measure it to within a trillionth of a second. A caesium clock loses one second every 300 years. Caesium clocks keep the world's official time and are carried on navigation satellites to pinpoint exact positions on Earth.

2. The ions are either in a low-energy or high-energy state.

Magnets

4. Inside, microwaves make the ions change into the higher-energy state.

6. The oscillator vibrates rhythmically due to an electric pulse, counting out time. The amount of microwaves in the chamber is linked to this oscillation.

Radiation

Quartz oscillator controls wavelength

3. A magnetic field deflects only the low-energy ions into a chamber.

Magnets

5. A detector counts the ions that emerge and sends a signal to a quartz oscillator.

The production of ions is linked to the oscillation in a feedback loop. If the oscillation slows down, there are fewer microwaves, and fewer energized caesium ions are detected. That results in an electric pulse restarting the quartz oscillator, boosting the number of ions at the detector. This feedback system ensures that the oscillator is always vibrating at the correct rate.

Detector

Feedback to oscillator
→

BARIUM

Atomic weight: 137.327
Colour: silver–grey
Phase: solid
Melting point: 729°C (1,344°F)
Boiling point: 1,897°C (3,447°F)
Crystal structure: body-centred cubic

Category: alkaline earth metal
Atomic number: 56

The name barium is derived from the Greek for 'heavy'. Barium's rocky compounds are dense and surprisingly weighty. The most frequently used compound is barium sulphate, known as baryte in its natural form. Baryte is used in dense drilling fluids and as a contrast medium for making stomach X-rays. Pure barium is a reactive metal, and so it is used in small amounts in high-temperature vacuum applications. Any oxygen molecules floating around would attack the components, and so barium is used to 'get' them before they can do any damage.

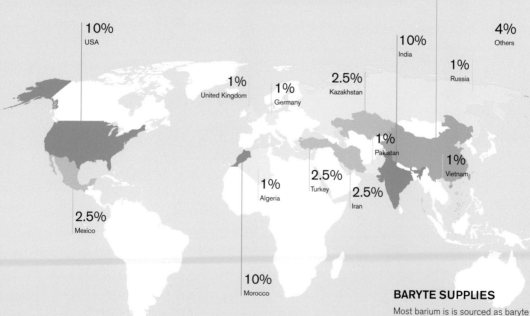

50%
China

10%
USA

10%
India

4%
Others

1%
Russia

1%
United Kingdom

1%
Germany

2.5%
Kazakhstan

1%
Pakistan

1%
Vietnam

1%
Algeria

2.5%
Turkey

2.5%
Iran

2.5%
Mexico

10%
Morocco

BARYTE SUPPLIES

Most barium is is sourced as baryte from mines all over the world. Witherite, a barium carbonate ore, is also used in smaller amounts.

LANTHANUM

Atomic weight: 138.90547
Colour: silver–white
Phase: solid
Melting point: 920°C (1,688°F)
Boiling point: 3,464°C (6,267°F)
Crystal structure: hexagonal

Category: lanthanide
Atomic number: 57

Lanthanum is a relatively common – although hard to source – heavy metal. About 70,000 tonnes is refined each year from its rare ores using a series of powerful displacement reactions. The metal has a growing number of uses in various smart materials. These are a new set of substances that have very specific properties, which vary in a precise way with changing conditions, such as temperature or electrical charge. The metal has a number of more prosaic uses in batteries, glassmaking and lighting.

GAS SPONGE

Lanthanum alloys are used as hydrogen sponges. The gas is absorbed by tiny spaces in the alloy, and the metals can squeeze in a huge amount of hydrogen – up to 400 times its own volume. These sponges are being developed as a means of storing hydrogen for use as a fuel.

TOP LENS

Lanthanum in lens glass reduces the aberration of images. All the light is focused to the same point instead of being slightly spread.

MAKING SPARKS

About a quarter of all refined lanthanum is used to make the flints that provide the spark for lighters.

H

GLOWING CLOTH

The cloth mantle of a gas lamp contains lanthanum oxide. This converts the heat of the burning gas into a bright, white light.

CERIUM

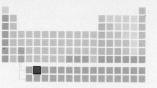

Atomic weight: 140.116
Colour: iron grey
Phase: solid
Melting point: 795°C (1,463°F)
Boiling point: 3,443°C (6,229°F)
Crystal structure: face-centred cubic

Category: lanthanide
Atomic number: 58

Cerium is the most common member of the lanthanide series, and it follows that it is most frequently used in technology. In common with many lanthanides, cerium is used in magnets and glasses, and as a catalyst.

8%
USA

LCD screen

4%
Australia

UV cut glass

1%
India

3%
Russia

HYBRID CAR

Cerium is a crucial material in modern cars, found in everything from the fuel to the dashboard touchscreen.

Fuel additives

Catalytic converter

84%
China

Hybrid battery

Glass and mirror polishing powder

PRASEODYMIUM

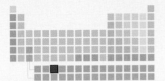

Atomic weight: 140.90765
Colour: silver–grey
Phase: solid
Melting point: 935°C (1,715°F)
Boiling point: 3,520°C (6,368°F)
Crystal structure: hexagonal

Category: lanthanide
Atomic number: 59

The name praseodymium means 'green twin' and refers to the way that the metal forms a flaky green oxide when left in the air. Many of the metal's applications are based around colour and light.

SLOW LIGHT

A silicate crystal doped with praseodymium has a remarkable effect on a beam of light shone through it. The speed of the beam slows from around 300 million metres per second to less than 1,000 metres per second. It is thought that this kind of slow-light technology could be used in faster and more efficient network switching to increase the fidelity of communications.

COLOURED TINTS

Praseodymium is used to create the dark tint in welder's goggles and shields. This filters out the intense light that would damage the eyes. Similarly, yellow glass lenses, which make it easier to see clearly in dim light conditions, contain this metal. Praseodymium also gives a green tinge to artificial gems, imitating the natural precious stone peridot.

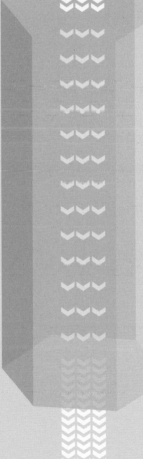

NEODYMIUM

60
Nd

Atomic weight: 144.242
Colour: silver–white
Phase: solid
Melting point: 1,024°C (1,875°F)
Boiling point: 3,074°C (5,565°F)
Crystal structure: hexagonal

Category: lanthanide
Atomic number: 60

Neodymium is alloyed with iron and boron to make NIB magnets. These are the most powerful magnets known, when comparing size to pulling power. A NIB magnet can lift:

1,000 TIMES
ITS OWN WEIGHT

SMALL BUT STRONG

The great power of NIB magnets means that they have been used to miniaturize many electromagnetic technologies, such as microphones, guitar pickups and other audio systems. Computer hard disks use a NIB magnet to read, write and erase data into a magnetic code.

GIVE IT A SPIN

Scaled up to a larger size, NIB magnets can produce immense torque in electric motors, which is why electric cars can accelerate so quickly.

PROMETHIUM

Atomic weight: 145
Colour: silver
Phase: solid
Melting point: 1,042°C (1,908°F)
Boiling point: 3,000°C (5,432°F)
Crystal structure: hexagonal

Category: lanthanide
Atomic number: 61

61
Pm

^{126}Pm   •   0.5s

^{145}Pm

17.7y

Promethium is too
radioactive to exist in
any useful quantity on
Earth. The only place
the element has been
detected in nature
is in the fireball of
a supernova. The
power of these
exploding stars
creates fresh supplies
of heavier elements.

11 x JELLYBEANS/
9 x PAPER CLIPS/
7 x PLAYING CARDS/
5 x PENNIES/
1 x AAA BATTERY/
3 x DICE

RAPID DECAY

Promethium has 38 isotopes and
none of them are stable. Most have
half-lives measured in minutes
or days. Pm-126 is the least stable,
while Pm-145 is the one that lasts
the longest.

THEORETICAL WEIGHT

Promethium is being formed all the time
by natural radioactivity on Earth, but
rapidly decays away into other elements.
Chemists have calculated that, at any
one instant, the Earth contains 12 grams
of the element, which is equivalent to:

SAMARIUM

Sm 62

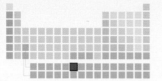

Atomic weight: 150.36
Colour: silver–white
Phase: solid
Melting point: 1,072°C (1,962°F)
Boiling point: 1,794°C (3,261°F)
Crystal structure: rhombohedral

Category: lanthanide
Atomic number: 62

Samarium and cobalt are combined to make magnets that are 10,000 times stronger in their effects than an equivalent iron magnet. Although this pulling power lags behind neodymium magnets, samarium magnets retain their magnetism better at high temperatures and so are used in more energy-intensive applications.

NO FUEL NEEDED

In July 2016, *Solar Impulse*, a one-seater aircraft, landed in Abu Dhabi, having flown around the world – in several stages – without the use of fuel. The aircraft's wings were lined with solar panels, which recharged batteries for flying at night. Power came from four super-efficient electric propeller engines. These used samarium magnets to create the required spinning motion.

A samarium-cobalt magnet is

10,000 times more powerful than an iron magnet

EUROPIUM

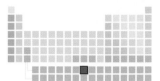

Atomic weight: 151.964
Colour: silver–white
Phase: solid
Melting point: 826°C (1,519°F)
Boiling point: 1,529°C (2,784°F)
Crystal structure: body-centred cubic

Category: lanthanide
Atomic number: 63

63

Eu

Despite being named after the continent of Europe, most of the world's europium reserves are in Asia (Mongolia) and North America (California). The metal's chief use is in the phosphors that glow inside light-emitting diodes (LEDs). These electronic components create the coloured pixels in flatscreen displays. Europium is involved in the red and blue LEDs. Fellow lanthanides terbium and ytterbium produce the green lights.

The richest source of europium is the Bayanobo Mine in Chinese Mongolia. The bastnäsite ore there contains a whopping

0.2%
europium

The Chinese mine is the world's main source of europium and produces a total of 45 per cent of all refined lanthanides.

SECRET SIGNS

The phosphorescent properties of europium mean it will glow under ultraviolet light. Many bank notes contain hidden symbols made with phosphorescent inks and implants, which are used to verify their authenticity. Although these security features are kept secret, it has been verified that the Euro notes do contain europium.

GADOLINIUM

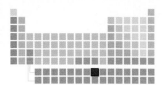

Atomic weight: 157.25
Colour: silver
Phase: solid
Melting point: 1,312°C (2,394°F)
Boiling point: 3,273°C (5,923°F)
Crystal structure: hexagonal

Category: lanthanide
Atomic number: 64

Gadolinium was the first element to be named directly after a human being. The metal was first extracted in 1886 from gadolinite, a dark, lustrous mineral that had been named by its discoverer – the Flemish chemist Johann Gadolin – after himself. So far, 19 elements have been named after people.

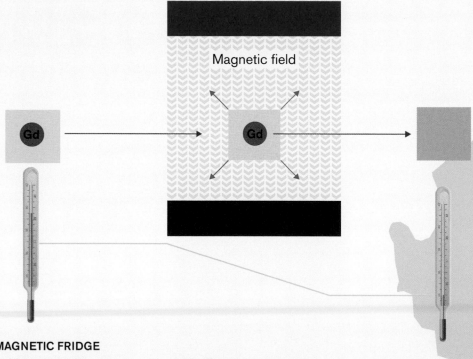

Magnetic field

MAGNETIC FRIDGE

Gadolinium is heated by a strong magnetic field. It sheds that heat immediately, and so the metal leaves the field colder than when it went in. This is a totally different system to the current methods of refrigeration, and may lead to cheaper and less polluting ways to keep things cold.

TERBIUM

Atomic weight: 158.92535
Colour: silvery white
Phase: solid
Melting point: 1,356°C (2,473°F)
Boiling point: 3,230°C (5,846°F)
Crystal structure: hexagonal

Category: lanthanide
Atomic number: 65

In keeping with its lanthanide neighbours, terbium has several niche applications. One of its more unusual properties is magnetostriction, where it vibrates in tune with an electric current running through it. That allows terbium to convert any flat surface – a table or window – into a loudspeaker. It transfers its vibrations to the surface, creating a sound wave. This effect is used in sonar systems already and may find many more applications.

PLACE NAME

Terbium was discovered in a mineral called yttria in 1843. Yttria is a complex mineral found at a Swedish mine, and named for the nearby town of Ytterby. Terbium was named for that town, and joined a unique set of metals – yttrium, erbium and ytterbium. No other place has so many elements named after it.

Y   Er   Yb

Tb

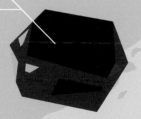

ELECTRIC YELLOW

Terbium phosphors glow a brilliant lemon yellow when electrified, and this light shines through a filter to produce the green pixels in flatscreen displays.

RIGHT PLACE, WRONG NAME

Gadolonite, the original source of gadolinium, was also found in Ytterby's prolific mine.

DYSPROSIUM

Dy 66

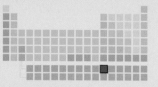

Atomic weight: 162.5
Colour: silvery white
Phase: solid
Melting point: 1,407°C (2,565°F)
Boiling point: 2,562°C (4,653°F)
Crystal structure: hexagonal

Category: lanthanide
Atomic number: 66

Dysprosium took several years of analysis to find among the jumble of lanthanide metals that cluster in 'rare earth' minerals. When it was finally shown to exist, it was named using the Greek phrase, 'hard to get'. The epithet remains true today, with little more than 100 tonnes produced each year.

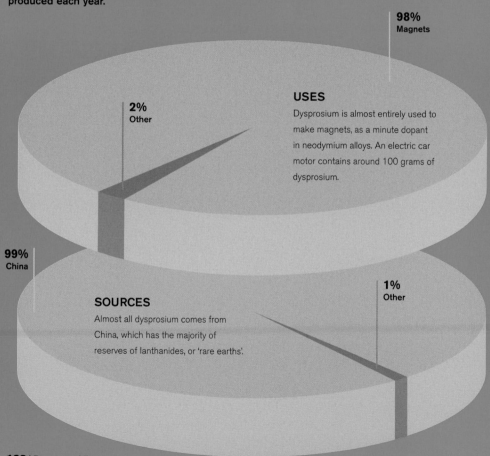

98%
Magnets

2%
Other

USES

Dysprosium is almost entirely used to make magnets, as a minute dopant in neodymium alloys. An electric car motor contains around 100 grams of dysprosium.

99%
China

1%
Other

SOURCES

Almost all dysprosium comes from China, which has the majority of reserves of lanthanides, or 'rare earths'.

HOLMIUM

Atomic weight: 164.93032
Colour: silvery white
Phase: solid
Melting point: 1,461°C (2,662°F)
Boiling point: 2,720°C (4,928°F)
Crystal structure: hexagonal

Category: lanthanide
Atomic number: 67

67
Ho

Holmium is named after the Swedish capital, Stockholm. The majority of lanthanides were discovered in strange minerals unearthed in Sweden. Holmium has a number of applications in electronics, but perhaps its most intriguing use is as a 'burnable poison' in the nuclear reactors powering submarines. While power plants use control rods to manage the fission, sub reactors are loaded with a 'poison' – usually holmium, boron and gadolinium – which soaks up the neutrons and keeps the reaction steady.

LASER KNIFE

Holmium is used to make the 2-μm laser used in surgery to cut or burn away tissue. The laser cuts more finely than a knife and can be targeted very precisely. It also sears the flesh, sealing the blood vessels.

COOL FORCE

Holmium has the largest magnetic moment of any element, which broadly means it produces the densest magnetic fields of all. However, this property only appears at 19K, or –254°C.

ERBIUM

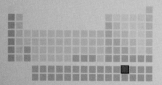

Atomic weight: 167.259
Colour: silver
Phase: solid
Melting point: 1,362°C (2,484°F)
Boiling point: 2,868°C (5,194°F)
Crystal structure: hexagonal

Category: lanthanide
Atomic number: 68

Another metal named for the Swedish town of Ytterby, erbium's parochial beginnings have now become truly global. The metal is used as a dopant in the laser amplifiers that keep telecommunication signals travelling along undersea cables.

DEGRADED SIGNALS

Even signals sent as flickering lasers weaken as they travel long distances through optical fibres. Every 70 kilometres or so, the signal gets a boost from an erbium amplifier.

LASER PUMP

The signal shines through a silica crystal doped with tiny amounts of erbium. The crystal has been energized by a laser, and it converts that into a fresh signal that is identical to the light that enters from the cable, but with a renewed power.

ROSE-TINTED

Pink glass is coloured using erbium oxide, which absorbs greener light and reflects reds and pale blues – making pink.

SHARK ATTACK

The amplifiers need a power supply, which runs through electric-cable bundles with the optical fibres. The electric field attracts sharks, which may attempt to bite the cables.

THULIUM

Atomic weight: 168.93421
Colour: silver–grey
Phase: solid
Melting point: 1,545°C (2,813°F)
Boiling point: 1, 950°C (3,542°F)
Crystal structure: hexagonal

Category: lanthanide
Atomic number: 65

69
Tm

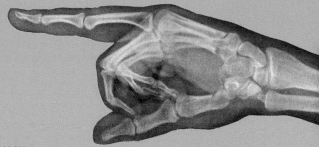

This grey metal is the grey sheep of the lanthanides. Although its main features do not stand apart from the rest of the series, thulium has no defining property or application. One of its radioactive isotopes is used as an X-ray source in portable medical scanners.

WHAT'S IN A NAME?

Thulium is named for Thule, a fabled land described in Greek myths that was located far to the north and was the source of the world's cold. No ancient explorers ever reached Thule, with most arriving in Scandinavia, where thulium was discovered in 1879 by Swede Per Teodor Cleve. Thulium was first isolated in 1911 by Briton Charles James. James had his work cut out; it took him 15,000 steps to finally purify the sample.

15,000

YTTERBIUM

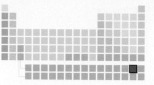

Atomic weight: 173.054
Colour: silver
Phase: solid
Melting point: 824°C (1,515°F)
Boiling point: 1,196°C (2,185°F)
Crystal structure: face-centred cubic

Category: lanthanide
Atomic number: 70

Ytterbium was the final element to be named after the Swedish town of Ytterby, after yttrium, terbium, and erbium. This may seem like a failure of imagination on the part of the chemists, but really it was the result of a long battle to lay claim to the discovery of the metal, which ran for 30 years.

1878 Jean Charles Galissard de Marignac isolates a mineral earth from erbium samples, which is called **YTTERBIA**.

1905 Carl Auer von Welsbach announces he has found two elements in ytterbia, which he names **ALDEBARANIUM** and **CASSIOPEIUM**.

1906 Charles James finds evidence of two new elements in ytterbia but does not offer names for them.

1907 Georges Urbain isolates two distinct compounds in ytterbia and calls them **NEOYTTERBIA** and **LUTECIA**.

1909 Urbain and Welsbach argue over the discovery, but chemists award the discovery to Urbain and settle on the names **YTTERBIUM** and **LUTETIUM**, although many German chemists prefer to use Welsbach's names until the 1950s.

1953 Pure **YTTERBIUM** is made for the first time.

50 tonnes

Annual production

$1,000 per kg

APPLYING PRESSURE

Ytterbium's conductivity varies with pressure, switching from a conductor to a semiconductor and back again as pressure rises. It is used to measure huge pressure, such as that in nuclear explosions and earthquakes.

LUTETIUM

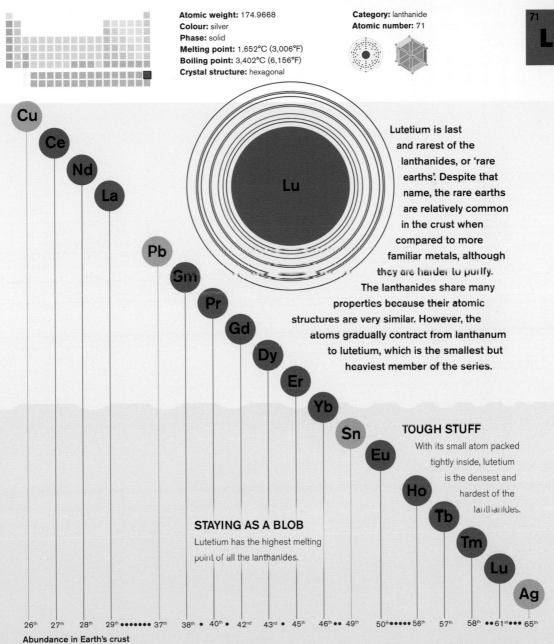

Atomic weight: 174.9668
Colour: silver
Phase: solid
Melting point: 1,652°C (3,006°F)
Boiling point: 3,402°C (6,156°F)
Crystal structure: hexagonal

Category: lanthanide
Atomic number: 71

71
Lu

Lu

Cu
Ce
Nd
La
Pb
Sm
Pr
Gd
Dy
Er
Yb
Sn
Eu
Ho
Tb
Tm
Lu
Ag

Lutetium is last and rarest of the lanthanides, or 'rare earths'. Despite that name, the rare earths are relatively common in the crust when compared to more familiar metals, although they are harder to purify. The lanthanides share many properties because their atomic structures are very similar. However, the atoms gradually contract from lanthanum to lutetium, which is the smallest but heaviest member of the series.

TOUGH STUFF

With its small atom packed tightly inside, lutetium is the densest and hardest of the lanthanides.

STAYING AS A BLOB

Lutetium has the highest melting point of all the lanthanides.

26th   27th   28th   29th •••••• 37th   38th • 40th • 42nd   43rd • 45th   46th •• 49th   50th ••••• 56th   57th   58th •• 61st ••• 65th

Abundance in Earth's crust

HAFNIUM

72

Hf

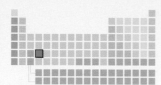

Atomic weight: 178.49
Colour: silver–grey
Phase: solid
Melting point: 2,233°C (4,051°F)
Boiling point: 4,603°C (8,317°F)
Crystal structure: hexagonal

Category: transition metal
Atomic number: 72

For a long time hafnium was a hidden element, with four per cent of zirconium samples actually being hafnium. The pair share almost identical chemical properties.

NEUTRON CAPTURE

Hafnium has five stable isotopes, and that makes it good at collecting neutrons in nuclear reactor control systems.

Hf-176 (5.3%)   Hf-178 (27.3%)   Hf-180 (35.1%)

Abundance of isotopes

Hf-174 (0.2%)   Hf-177 (18.6%)   Hf-179 (13.5%)

SUPPLY AND DEMAND

About 80 tonnes of pure hafnium is made each year. The price has recently increased, with a rise in demand for new nuclear power projects.

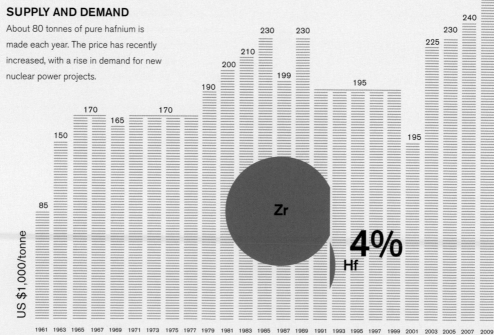

US $1,000/tonne

85   150   170   165   170   190   200   210   230   199   230   195   195   225   230   240

Zr

4%
Hf

1961 1963 1965 1967 1969 1971 1973 1975 1977 1979 1981 1983 1985 1987 1989 1991 1993 1995 1997 1999 2001 2003 2005 2007 2009

TANTALUM

Chart scale: 0, 200, 400, 600, 800, 1,000, 1,200, 1,400, 1,600

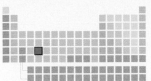

Atomic weight: 180.9479
Colour: silvery grey
Phase: solid
Melting point: 3,017°C (5,463°F)
Boiling point: 5,458°C (9,856°F)
Crystal structure: body-centred cubic

Category: transition metal
Atomic number: 73

73
Ta

Tantalum is used to make the tiny capacitors (charge stores) used in electronic gadgets, such as phones and tablets. One of its most significant ores is coltan, which also contains the metal niobium. Coltan is found mostly in central Africa. This region is likely to become a leading producer of tantalum in the future.

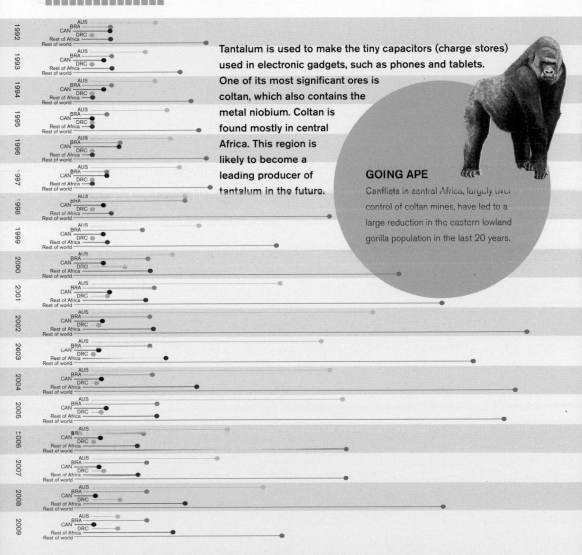

GOING APE

Conflicts in central Africa, largely over control of coltan mines, have led to a large reduction in the eastern lowland gorilla population in the last 20 years.

Years (left axis): 1991, 1992, 1993, 1994, 1995, 1996, 1997, 1998, 1999, 2000, 2001, 2002, 2003, 2004, 2005, 2006, 2007, 2008, 2009

Categories per year: AUS, BRA, CAN, DRC, Rest of Africa, Rest of world

TUNGSTEN

74
W

The word tungsten comes from the Swedish for 'heavy stone'. The metal was discovered in a dense mineral called wolframite (meaning 'wolf dirt'); the 'W' symbol comes from that. Tungsten is most visible as the glowing filament in an incandescent light bulb, but that is only a minor use.

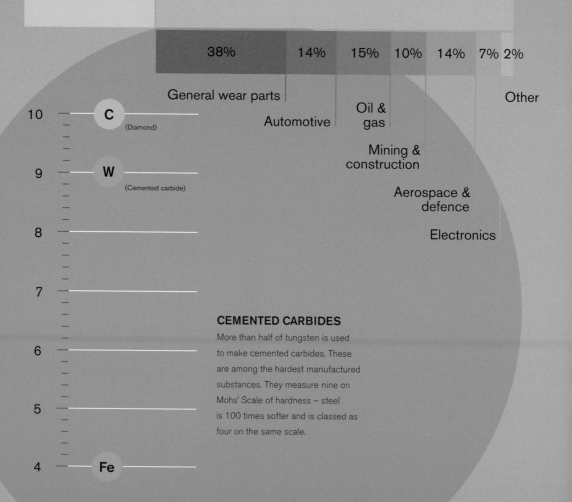

20%
Steels/alloys

55%
Cemented carbide

38% 14% 15% 10% 14% 7% 2%

General wear parts

Automotive

Oil & gas

Mining & construction

Aerospace & defence

Electronics

Other

10 — C
(Diamond)

9 — W
(Cemented carbide)

8

7

6

5

4 — Fe

CEMENTED CARBIDES

More than half of tungsten is used to make cemented carbides. These are among the hardest manufactured substances. They measure nine on Mohs' Scale of hardness – steel is 100 times softer and is classed as four on the same scale.

Atomic weight: 183.84
Colour: silvery white
Phase: solid
Melting point: 3,422°C (6,192°F)
Boiling point: 5,555°C (10,031°F)
Crystal structure: body-centred cubic

Category: transition metal
Atomic number: 74

SOURCES

Around 75,000 tonnes of tungsten
are smelted from ores each year.
About 80 per cent of that comes
from China.

17%
Mill products

8%
Other

4%

MELTING POINT

Tungsten has the highest melting point
of any metal. Only carbon stays solid at
higher temperatures. An oxyacetylene
torch is hot enough to melt it, but
a large sunspot, a cool zone
on the Sun, may not be!

C

Lightbulbs

THERMAL EXPANSION

Pure tungsten expands only slowly
when heated. With every increase of one
degree of temperature, steel expands
three times more than tungsten.

3,642°C

3,480°C

W

3,422°C

3,000°C

W

Fe

RHENIUM

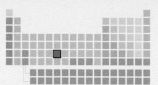

75

Re

Atomic weight: 186.207
Colour: silvery white
Phase: solid
Melting point: 3,186°C (5,767°F)
Boiling point: 5,596°C (10,105°F)
Crystal structure: hexagonal

Category: transition metal
Atomic number: 75

Rhenium, discovered in 1925, was the last stable element to be found. Less than 50 tonnes are refined each year, mostly from manganese and molybdenum ores. However, the annual demand for the metal is around 60 tonnes, and the extra ten tonnes comes from recycling.

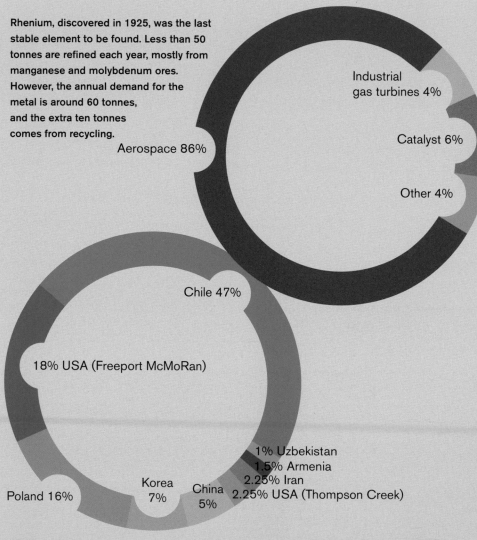

Industrial gas turbines 4%

Catalyst 6%

Other 4%

Aerospace 86%

Chile 47%

18% USA (Freeport McMoRan)

Poland 16%

Korea 7%

China 5%

1% Uzbekistan
1.5% Armenia
2.25% Iran
2.25% USA (Thompson Creek)

OSMIUM

Atomic weight: 190.23
Colour: bluish grey
Phase: solid
Melting point: 3,033°C (5,491°F)
Boiling point: 5,012°C (9,054°F)
Crystal structure: hexagonal

Category: transition metal
Atomic number: 76

76

Os

Osmium is the rarest element in the Earth's crust, with only one atom of it per ten billion other atoms in rocks. It is also the densest (although some measures give iridium that prize).

CURRENT USES

Osmium bonds to oils and fats well. It is used to stain biological samples for viewing with electron microscopes, and powdered osmium clings to oily residues left by fingerprints.

A hard osmium-iridium alloy is used in the nibs of fountain pens.

Osmium =
1 atom

Fingerprints

Others = 10 billion
atoms

SUPERSEDED ROLES

Osmium was the metal of choice for early lightbulbs before tungsten took over. It was also used to make the stylus for 1950s gramophones.

IRIDIUM

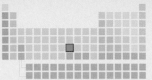

77

Ir

Atomic weight: 192.217
Colour: silvery white
Phase: solid
Melting point: 2,466°C (4,471°F)
Boiling point: 4,428°C (8,002°F)
Crystal structure: face-centred cubic

Category: transition metal
Atomic number: 77

Iridium is almost universally rare in the Earth's rocks. However, a thin layer of quartz dust that is found in rocks all over the world contains an unusually high level of the metal. The dust dates from 65 million years ago, a point in history when a ten-kilometre-wide asteroid smashed into Mexico. The debris of this impact shrouded Earth in a cloud of dust, which eventually settled to form the layer we see today. The iridium in it is from the exploded space rock. Biologists theorize that this global event was what led to the extinction of the dinosaurs.

PLATINUM

Atomic weight: 195.078
Colour: silvery white
Phase: solid
Melting point: 1,768°C (3,215°F)
Boiling point: 3,825°C (6,917°F)
Crystal structure: face-centred cubic

Category: transition metal
Atomic number: 78

78

Pt

35.9%
Jewellery

40.4%
Autocatalyst

6.4%
Investment

5.9%
Chemical

3.2%
Medical &
biomedical

3%
Glass

2.6%
Petroleum

2.6%
Electrical

Platinum, from the Spanish for 'little silver', is not just used for jewellery. In fact, 60 per cent has an industrial, not cosmetic, function. About 80 per cent of the 175 tonnes of platinum produced each year comes from southern Africa.

1,500–7,000kg

1,500–7,000kg

6–1,500kg

6–1,500kg

10,000–30,000kg

10,000–30,000kg

6–1,500kg

6–1,500kg

6–1,500kg

10,000–30,000kg

30,000–133,000kg

6–1,500kg

GOLD

79
Au

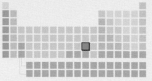

Gold is the epitome of value. For millennia, the uniquely yellowish metal has been a symbol of wealth. Almost entirely inert, the metal is found pure in nature and always remains pristine and corrosion-free. Gold is a safe bet. Unlike other elements, it will never fade or crumble to dust.

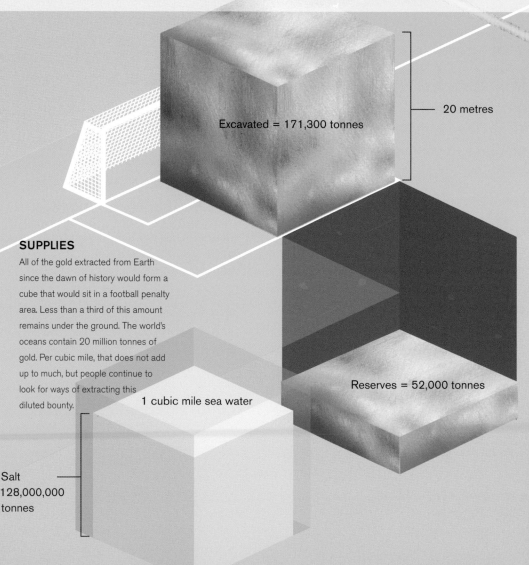

Excavated = 171,300 tonnes

20 metres

SUPPLIES

All of the gold extracted from Earth since the dawn of history would form a cube that would sit in a football penalty area. Less than a third of this amount remains under the ground. The world's oceans contain 20 million tonnes of gold. Per cubic mile, that does not add up to much, but people continue to look for ways of extracting this diluted bounty.

Reserves = 52,000 tonnes

1 cubic mile sea water

Salt
128,000,000
tonnes

Gold
17 kilograms

Atomic weight: 196.96655
Colour: metallic yellow
Phase: solid
Melting point: 1,064°C (1,948°F)
Boiling point: 2,856°C (5,173°F)
Crystal structure: face-centred cubic

Category: transition metal
Atomic number: 79

1g

MATERIAL PROPERTIES

Gold is the most ductile element, meaning it can be pulled into wires without breaking. It is also the most malleable, and can be hammered flat into foils so thin that they become transparent.

165m

1m²

DUST AND NUGGETS

Gold is generally sourced as dust that is extracted by crushing rocks. Occasionally larger masses, or nuggets, are found. The largest is the Welcome Stranger, which was discovered in Australia in 1869.

Weight	97.14 kilograms
Dimensions	61 x 31 centimetres

MERCURY

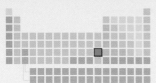

Atomic weight: 200.59
Colour: silvery white
Phase: liquid
Melting point: -39°C (-38°F)
Boiling point: 357°C (674°F)
Crystal structure: rhombohedral

Category: transition metal
Atomic number: 80

Mercury is one of only two elements that is liquid in standard conditions. It is named after the messenger of the Roman gods, who was fast moving and hard to handle. An earlier name was quicksilver, and the 'Hg' symbol is derived from the Latin term *hydrargyrum*, or 'water silver'.

DANGEROUS STUFF!

Breathing in the vapour of mercury leads to permanent damage to the nervous system. As a result, it is rarely used in modern manufacturing. However, mercury is still being released by certain industries, especially wildcat gold mining, where gold is extracted from rock by dissolving it in mercury.

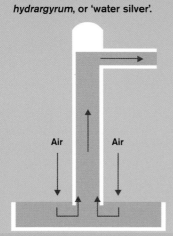

Air   Air

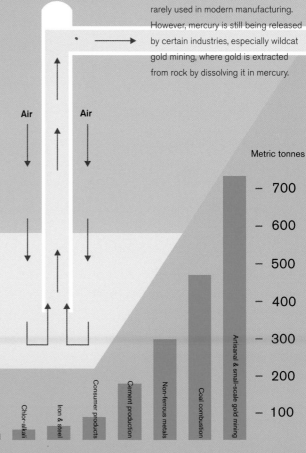

Air   Air

HIGH DENSITY

Mercury is 14 times denser than water. In the 17th century, engineers found they could not siphon water higher than around ten metres. They modelled the problem on a smaller scale using mercury, and found that it could only rise to 76 centimetres, a 14th of the water height. It found that these heights were governed by air pressure, the weight of the atmosphere pushing on the liquid. This was the beginnings of the barometer, a device for measuring gas pressures, which were the first clues to the atomic nature of the elements.

Metric tonnes

- 700
- 600
- 500
- 400
- 300
- 200
- 100

Oil refining
Chlor-alkali
Iron & steel
Consumer products
Cement production
Non-ferrous metals
Coal combustion
Artisanal & small-scale gold mining

THALLIUM

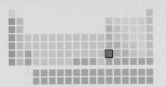

Atomic weight: 204.3833
Colour: silvery white
Phase: solid
Melting point: 304°C (579°F)
Boiling point: 1,473°C (2,683°F)
Crystal structure: hexagonal

Category: post-transition metal
Atomic number: 81

81
Tl

Thallium is a toxic heavy metal, and exposure causes effects all over the body – and eventually an awful death. Some call thallium sulphate the 'poisoner's poison', because it is colourless, tasteless and hard to detect in the body.

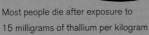 Constipation

 Pain in extremities

 Vomiting and nausea

 Abdominal pain

 'Mees' lines on nails

 Hair loss

 Increased heart rate

 Convulsions, coma and death

15mg/kg

MEDIAN DOSE

Most people die after exposure to 15 milligrams of thallium per kilogram of body weight.

SHINE A LIGHT

Thallium is named after a leaf-green colour in its emission spectrum. Thallium poisoning is detected by shining light through the sufferer's urine. Any thallium in it will absorb the green light.

Light ⟶

LEAD

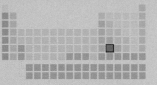

82
Pb

Atomic weight: 207.2
Colour: grey
Phase: solid
Melting point: 327°C (621°F)
Boiling point: 1,749°C (3,180°F)
Crystal structure: face-centred cubic

Category: post-transition metal
Atomic number: 82

 Blurred vision

 Tingling extremities

 Slurred speech

 Constipation and diarrhoea

 Memory loss

 Kidney failure

 Convulsions

 Livid (blue–grey) skin

 Hearing loss

 Anaemia

 Infertility

 General fatigue

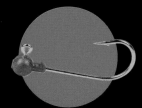

Fishing sinkers

Solder

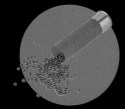

Shotgun pellets

Lead was probably the first metal to be refined in large amounts, dating back 9,000 years. However, in all that time, the metal has been poisoning people. Lead seldom kills, but it does cause a range of chronic disorders of the gut and the nervous system. Traditional applications of lead have been steadily phased out over the last 40 years.

BISMUTH

Atomic weight: 208.98040
Colour: silver
Phase: solid
Melting point: 272°C (521°F)
Boiling point: 1,564°C (2,847°F)
Crystal structure: trigonal

Category: post-transition metal
Atomic number: 83

83
Bi

Bismuth is not included in the list of radioactive elements, but its atoms do decay into thallium – very slowly. The element's half-life is a billion times longer than the current age of the Universe.

= x 1,000,000,000

Replaced by

Fishing sinkers

Solder

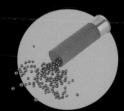

Shotgun pellets

Pepto-Bismol

LEAD REPLACEMENT

Bismuth is almost as dense as lead and has a low melting point. As a result, it is a good substitute for some applications that once used lead. Another widespread use of the metal is in a general stomach remedy for curing indigestion, soothing stomach aches and tackling diarrhoea.

POLONIUM

84
Po

Atomic weight: 209
Colour: silvery grey
Phase: solid
Melting point: 254°C (489°F)
Boiling point: 962°C (1,764°F)
Crystal structure: cubic

Category: metalloid
Atomic number: 84

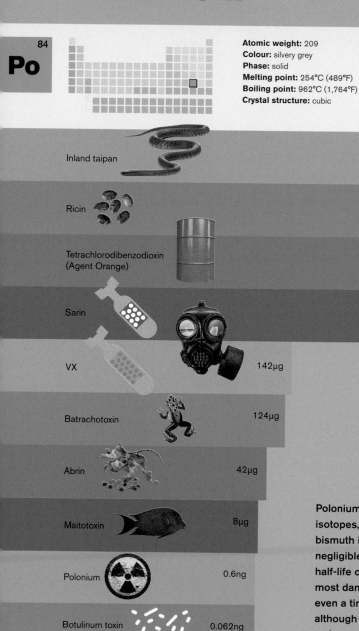

Inland taipan   1,500µg per kg of body weight ⟶

Ricin   1,300µg

Tetrachlorodibenzodioxin
(Agent Orange)   1,200µg

Sarin   1,000µg

VX   142µg

Batrachotoxin   124µg

Abrin   42µg

Maitotoxin   8µg

Polonium   0.6ng

Botulinum toxin   0.062ng

Polonium is the first element with only radioactive isotopes, if the ultra-slow and negligible decay of bismuth is ignored. Polonium's decay is far from negligible. Its most common isotope, Po-210, has a half-life of just 138 days. It emits alpha particles, the most damaging form of radiation, and so ingesting even a tiny quantity of Po-210 is invariably fatal – although it takes a few weeks to die. Despite being a slow-acting killer, polonium's toxicity is second only to botulin in a list of deadly chemicals.

ASTATINE

Atomic weight: 210
Phase: solid
Colour: unknown
Melting point: 302°C (576°F)
Boiling point: 337°C (639°F)
Crystal structure: unknown

Category: halogen
Atomic number: 85

85
At

5,974,000,000,000,000,000,000,000kg

Astatine is the fifth halogen, but it is
so impossibly radioactive that only tiny
amounts exist on Earth. Astatine atoms
are always being produced by the decay
of other radioactive elements. However, the
most stable astatine isotope has a half-life of just seven hours –
and most of the others decay away in minutes – so these atoms
do not stay around for long.

At   30g

RADON

86

Rn

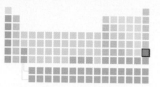

Atomic weight: 222
Colour: colourless
Phase: gas
Melting point: -71°C (-96°F)
Boiling point: -62°C (-79°F)
Crystal structure: n/a

Category: noble gas
Atomic number: 86

Radon is a noble gas and so is not active in chemical reactions. However, it is very radioactive and poses the largest risk to health from exposure to natural radioactivity. Being a gas means that radon can escape from the rocks where it is formed by natural decay. Denser than air, radon can gather in dangerous concentrations in cellars and houses.

GAS AREAS

Radon is most common in regions with granite bedrock. Once detected, fans and vents are used to disperse the radon from homes.

	A		B		C		D		picoCuries per litre of air
0.2		0.7		1.4		2.7		5	

FRANCIUM

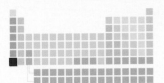

Atomic weight: 223
(isotope francium-223)
Phase: solid
Colour: unknown
Melting point: 27°C (80°F) (estimated)
Boiling point: 680°C (1,256°F) (estimated)

Category: alkali metal
Atomic number: 87

Radioactive francium atoms only exist fleetingly in the Earth's crust. It was the last element to be discovered in nature, by French chemist Marguerite Perey, in 1939. She named it after France, making it the second element to get that honour after gallium.

CAUGHT IN A TRAP

Francium is made in the laboratory by bombarding gold with oxygen ions. The largest quantity of francium ever isolated was a cluster of 300,000 atoms, held in a magnetic trap. This sample was still truly tiny when compared with the number of atoms that would be required to make a cubic centimetre of francium.

300,000 atoms

1cm³ ——— 10,000,000,000,000,000 atoms

RADIUM

88	
Ra	

Atomic weight: 226
Colour: white
Phase: solid
Melting point: 700°C (1,292°F)
Boiling point: 1,737°C (3,159°F)
Crystal structure: body-centred cubic

Category: alkaline earth metal
Atomic number: 88

1,600 years

XII
XI
X
IX

DEATH WATCH

Luminous radium paint was used to make clocks and watches glow in the dark. With a half-life of 1,600 years, these toxic time pieces will not fade for centuries.

Radium captured the public imagination when it was discovered in 1898. The soft, green glow emitted by its radioactive compounds was seen as a sign of a restorative power. However, a generation later, radium's dangers to health had become well documented.

CURE FOR EVERYTHING

In the early 20th century, radioactivity was sold as a panacea. Radium-infused water and baths salts were said to boost energy, radium creams combatted the signs of ageing, while toothpastes whitened teeth.

BAD THERAPY

By the 1920s it was becoming apparent that radium was causing cancers, especially in bones where the metal replaced natural calcium. However, many people swore by radium treatments well into the 1950s.

ACTINIUM

Atomic weight: 227
Colour: silver
Phase: solid
Melting point: 1,050°C (1,922°F)
Boiling point: 3,198°C (5,788°F)
Crystal structure: face-centred cubic

Category: actinide
Atomic number: 89

89

Ac

This silvery, dense metal is the first member of the actinide series, which is named after it in keeping with the system set by lanthanum, its upper neighbour in the periodic table. The actinides are all radioactive and include the heaviest natural elements of all.

ALPHA SOURCE

Looking somewhat nondescript in daylight, actinium takes on a deep-blue glow in the dark. This light is from the emission of alpha particles.

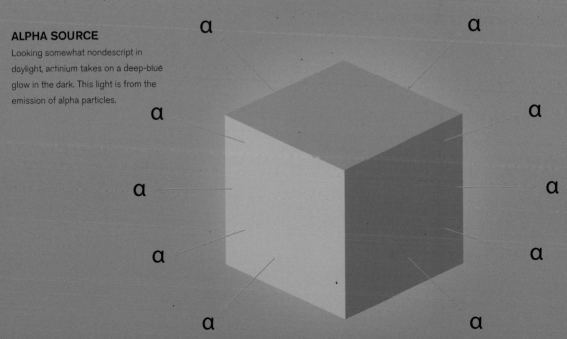

α α

α α

α α

α α

α α

THORIUM

Atomic weight: 232.0381
Colour: silver
Phase: solid
Melting point: 1,842°C (3,348°F)
Boiling point: 4,788°C (8,650°F)
Crystal structure: face-centred cubic

Category: actinide
Atomic number: 90

Thorium is the most common radioactive element on Earth. It has a series of niche applications, especially making heat-resistant glasses and alloys. It is refined from monazite, a phosphate ore.

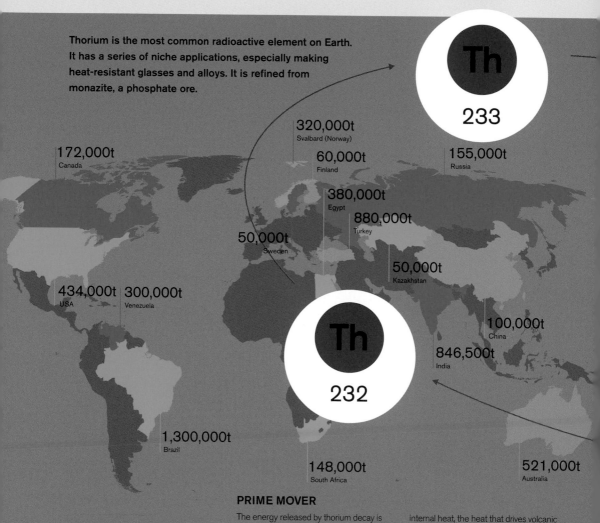

Th 233

Th 232

320,000t
Svalbard (Norway)

60,000t
Finland

172,000t
Canada

155,000t
Russia

380,000t
Egypt

880,000t
Turkey

50,000t
Sweden

50,000t
Kazakhstan

434,000t
USA

300,000t
Venezuela

100,000t
China

846,500t
India

1,300,000t
Brazil

148,000t
South Africa

521,000t
Australia

PRIME MOVER

The energy released by thorium decay is the single-largest contributor to the planet's internal heat, the heat that drives volcanic activity and tectonic plate movements.

PROTACTINIUM

Atomic weight: 231.03588
Colour: silver
Phase: solid
Melting point: 1,568°C (2,854°F)
Boiling point: 4,027°C (7,280°F)
Crystal structure: tetragonal

Category: actinide
Atomic number: 91

91
Pa

Protactinium is found in incredibly small amounts in nature, mostly inside uranium ores. Uranium decays into protactinium, which can then form actinium – hence this element's name, which means 'before actinium'.

THORIUM CYCLE

A few isotopes of thorium can undergo nuclear fission, releasing large amounts of heat. However, these isotopes are very rare, which makes thorium impractical for use as a regular nuclear fuel. Nevertheless, it may be possible to use thorium to drive nuclear fission in power plants using the 'thorium fuel cycle'. The thorium-232 isotope will absorb a neutron, forming thorium-233. This decays into protactinium-233, which in turn decays into uranium-233. This synthetic uranium isotope is fissile and can be used as a nuclear fuel. As it splits it releases neutrons, which are collected by thorium-232 – and the cycle begins again.

Pa
233

U
233

Neutron

URANIUM

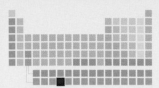

Atomic weight: 238.02891
Colour: silver–grey
Phase: solid
Melting point: 1,132°C (2,070°F)
Boiling point: 4,131°C (7,468°F)
Crystal structure: orthorhombic

Category: actinide
Atomic number: 92

Uranium is the most familiar radioactive element, and it was the first to be discovered. It was first identified in 1788, and it was uranium ore that gave the first evidence of radioactivity in 1896. Many of the remaining radioactive elements are made when uranium decays, and were discovered through the analysis of uranium ore.

U-238

More than 99 per cent of uranium is U-238, which has a half-life of 4.5 billion years. That means only half of the uranium present when the Earth formed still exists.

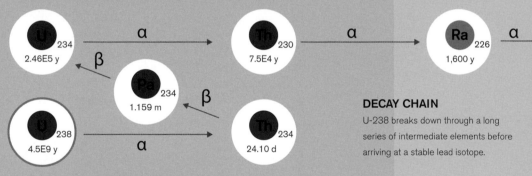

DECAY CHAIN

U-238 breaks down through a long series of intermediate elements before arriving at a stable lead isotope.

ARMED AND DANGEROUS

About 0.7 per cent of uranium is the U-235 isotope. This is the fissile atom that can produce the fission chain reaction. Uranium is refined around the world to access this isotope, which is used as a heat source in nuclear power plants and in nuclear weapons. Non-fissile uranium isotopes (although still radioactive) are used in military armour.

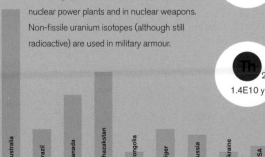

Australia	Brazil	Canada	Khazakstan	Mongolia	Niger	Russia	Ukraine	USA	Uzbekistan
28%	6%	12%	16%	2%	6%	5%	2%	3%	2%

THORIUM CHAIN

Th-232 has a half-life of 14 billion years. Its decay chain also leads to a stable lead isotope.

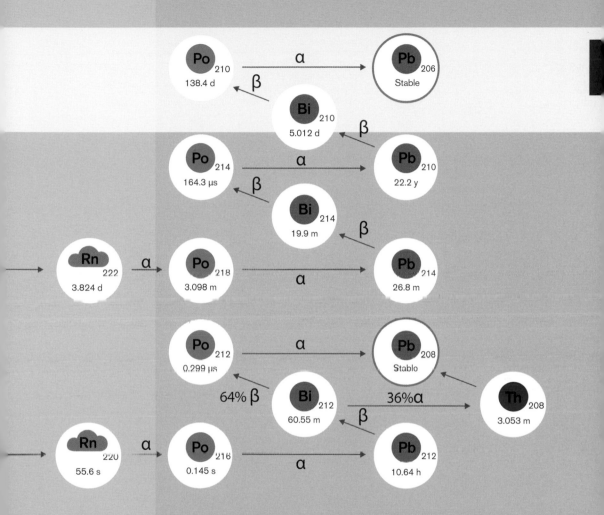

ELEMENTAL MIX

Uranium and thorium ores, such as pitchblende and monazite, contain trace amounts of several other radioactive elements. Radium and radon are more soluble than the others and may be washed out and escape into the wider environment.

NEPTUNIUM

93
Np

Atomic weight: 237
Colour: silver
Phase: solid
Melting point: 637°C (1,179°F)
Boiling point: 4,000°C (7,232°F)
Crystal structure: orthorhombic

Category: actinide
Atomic number: 93

Np

Neptunium was the first transuranic element, meaning it was the first element to be found that is heaver than uranium, the largest natural element. It was discovered in 1940 as part of the research and development of nuclear reactors. Since then, minute amounts of neptunium have been found in the decay chains of rare uranium isotopes.

PLANET PLAN

Uranium was named for the planet Uranus, which had been discovered shortly before the metal was first described. Therefore, when element 93 was found, it was named for the next planet along, Neptune. Element 94 followed soon after, and was named for Pluto, then still regarded as a planet.

 **URANUS**

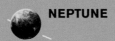 **NEPTUNE**

ALL GONE NOW

The half-life of the most stable neptunium isotope is two million years. That means that any neptunium that existed on the young Earth would have decayed away to almost nothing within 80 million years.

 PLUTO

80,000,000 years

PLUTONIUM

Atomic weight: 244
Colour: silver–white
Phase: solid
Melting point: 639°C (1,183°F)
Boiling point: 3,228°C (5,842°F)
Crystal structure: monoclinic

Category: actinide
Atomic number: 94

Plutonium was discovered during the Manhattan Project, which developed nuclear weapons in the Second World War. It was made by bombarding uranium with radiation, which added mass to the atoms. Many of the isotopes made this way were quite stable, with half-lives measured in thousands of years. One isotope, Pu-239, was fissile, and was used to make the first nuclear bomb.

BIG BANGS

The Gadget was the first nuclear bomb, exploded in the Trinity Test in Arizona in 1945. It used 6.4 kilograms of plutonium, and was almost identical to the Fat Man bomb used over Nagasaki, Japan, shortly afterwards. The Hiroshima Little Boy bomb, dropped in between, used uranium and was less powerful. However, the power of these early bombs is minute when compared with the thermonuclear weapons filling today's arsenals. The Tsar Bomba, a Russian thermonuclear device, was the largest artificial explosion in history.

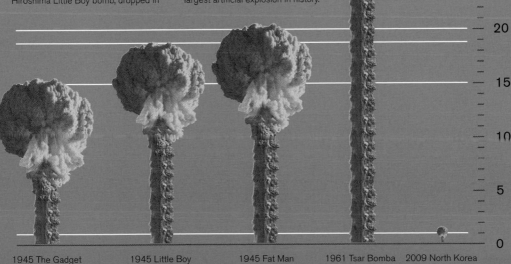

Kt
57,000

56,995

25

20

15

10

5

0

1945 The Gadget   1945 Little Boy   1945 Fat Man   1961 Tsar Bomba   2009 North Korea

AMERICIUM

95

Am

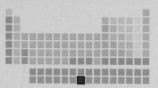

Atomic weight: 243
Colour: silver–white
Phase: solid
Melting point: 1,176°C (2,149°F)
Boiling point: 2,607°C (4,725°F)
Crystal structure: hexagonal

Category: actinide
Atomic number: 95

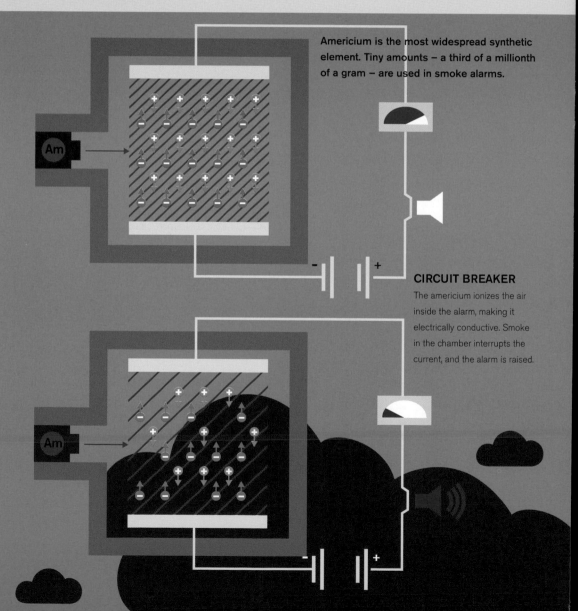

Americium is the most widespread synthetic element. Tiny amounts – a third of a millionth of a gram – are used in smoke alarms.

CIRCUIT BREAKER

The americium ionizes the air inside the alarm, making it electrically conductive. Smoke in the chamber interrupts the current, and the alarm is raised.

CURIUM

Atomic weight: 247
Colour: silver
Phase: solid
Melting point: 1,340°C (2,444°F)
Boiling point: 3,110°C (5,630°F)
Crystal structure: hexagonal close-packed

Category: actinide
Atomic number: 96

96

Cm

Curium is a strong alpha emitter. Most of its isotopes release these large particles as they decay. As a result, curium is an essential part of today's space-exploration probes, including all the Mars rovers and the Philae comet lander. Curium's radiation is used to excite rock samples on alien worlds. The light sent back tells the probe what the samples are made of.

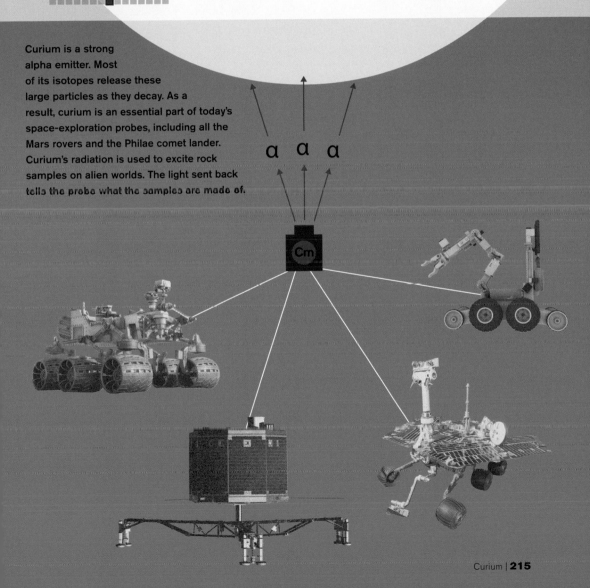

BERKELIUM

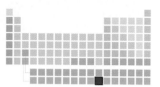

97
Bk

Atomic weight: 247
Colour: silver
Phase: solid
Melting point: 986°C (1,807°F)
Boiling point: unknown
Crystal structure: hexagonal close-packed

Category: actinide
Atomic number: 97

Berkelium was discovered at a time when new synthetic elements were being made every few years. Finding new names was a problem, and so for elements 95 to 97, inspiration was drawn from the period above.

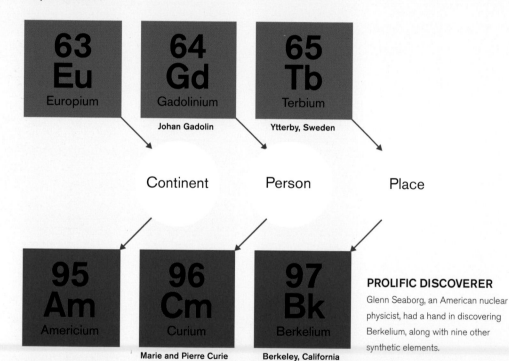

63
Eu
Europium

64
Gd
Gadolinium

65
Tb
Terbium

Johan Gadolin

Ytterby, Sweden

Continent

Person

Place

95
Am
Americium

96
Cm
Curium

97
Bk
Berkelium

Marie and Pierre Curie

Berkeley, California

PROLIFIC DISCOVERER

Glenn Seaborg, an American nuclear physicist, had a hand in discovering Berkelium, along with nine other synthetic elements.

93
Np
Neptunium

94
Pu
Plutonium

95
Am
Americium

96
Cm
Curium

97
Bk
Berkelium

98
Cf
Californium

99
Es
Einsteinium

100
Fm
Fermium

101
Md
Mendelevium

102
No
Nobelium

CALIFORNIUM

Atomic weight: 251
Colour: silver–white
Phase: solid
Melting point: 900°C (1,652°F)
Boiling point: 1,745°C (3,173°F)
(estimated)
Crystal structure: double hexagonal

Category: actinide
Atomic number: 98

98

Cf

Californium is the raw material used to make many of the larger, transfermium synthetic elements. It is also a superlative neutron source; one microgram releases 139 million neutrons a minute. As a result, this element is used to ignite nuclear fuels and is used in medical scanners and treatments that require neutrons.

$27,000,000

1kg

PRICE POINT

In all its applications, californium is used in tiny amounts because it is the most expensive element of all.

EINSTEINIUM

99

Es

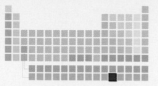

Atomic weight: 252
Colour: silver
Phase: solid
Melting point: 860°C (1,580°F)
Boiling point: unknown
Crystal structure: face-centred cubic

Category: actinide
Atomic number: 99

Einsteinium, named after the great physicist, was discovered in the remains of the Ivy Mike bomb test in 1952. This bomb was the first thermonuclear weapon, otherwise known as an H-bomb. Instead of sourcing its explosive power from nuclear fission, an H-bomb releases even more energy from the fusion of radioactive hydrogen. Nevertheless, a small fission explosive is used to detonate the fusion process.

Es
50mg

FUSION POWER

Einsteinium was created by Ivy Mike when uranium atoms in the bomb absorbed dozens of neutrons to make californium-253. This decayed into einsteinium. Despite its immense explosive power, Ivy Mike produced just 50 milligrams of einsteinium.

SEEING IS BELIEVING

Einsteinium is the heaviest element that can be made in amounts large enough to be seen with the naked eye (just).

10 megatonne explosion

Number of atoms produced

FERMIUM

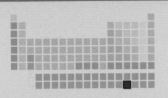

Atomic weight: 257
Colour: unknown
Phase: solid
Melting point: 1,527°C (2,781°F)
Boiling point: unknown
Crystal structure: unknown

Category: actinide
Atomic number: 100

100
Fm

Named for Enrico Fermi, the Italian physicist who first ran a controlled fission chain reaction, thus starting the race for nuclear weapons and nuclear power, fermium was first seen in the aftermath of the Ivy Mike H bomb test in 1952. There are 19 isotopes of fermium, of which the most stable form has a half-life of just 100 days.

DECREASING YIELDS

Fermium is the largest transuranic element that is made by nuclear explosions. The yields of the elements steadily decrease with atomic number, but isotopes with even numbers of nuclear particles, which form into stable pairs, are more likely to be produced.

10^{22}

10^{21}

10^{20}

10^{19}

10^{18}

10^{17}

10^{16}

10^{15}

10^{14}

Mass number

| 240 | 245 | 250 | 255

TRANSFERMIUM ELEMENTS

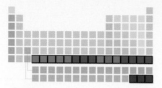

Elements with atomic numbers greater than fermium (100) fall into two groups. Elements 101 to 103 complete the actinide series. The remaining elements fill up the seventh period of the periodic table, and are known as 'superheavy' elements.

MENDELEVIUM (101)

Named for Dmitri Mendeleev, the Russian inventor of the periodic table. One isotope, Md-250, spontaneously splits in two, rather than decay in the normal way.

NOBELIUM (102)

Named after Alfred Nobel, the Swedish chemist, explosives magnate and philanthropist. The first element to have a half-life shorter than one hour.

LAWRENCIUM (103)

Named in honour of Ernest Lawrence, the American inventor of the particle accelerator, which became crucial in making synthetic elements.

BOHRIUM (107)

Named after Niels Bohr, a leading figure in the foundation of quantum physics. The most stable isotope of this element has a half-life of just 61 seconds.

HASSIUM (108)

Named after the German state of Hesse, where it was first made. This metal has a half-life of 30 seconds.

MEITNERIUM (109)

Predicted to be the densest substance on Earth, only a few atoms can be made at a time. It is only the second element to be named after a woman, Lise Meitner, the co-discoverer of nuclear fission.

NIHONIUM (113)

Named after Japan, where it was synthesized in 2004. It is a member of Group 3, although its physical and chemical properties are uncertain.

FLEROVIUM (114)

This element reacts with gold and is the largest element to have been made into a compound. It is named for Russian physicist Georgy Flyorov.

MOSCOVIUM (115)

No isotopes of this element, which is named after the Russian capital, have half-lives longer than a second.

TRANSFERMIUM WARS

Naming the first set of superheavy elements became a fraught political battle between the United States and the Soviet Union, whose scientists disagreed on whom had discovered them first. Disagreement raged for 35 years until 1997, when elements 104 to 109 were finally given internationally accepted names.

RUTHERFORDIUM (104)

The first superheavy element included in the transition series is named after Ernest Rutherford, the New Zealand scientist who discovered the atomic nucleus in 1911.

DUBNIUM (105)

Named after Dubna, the Russian city near Moscow that is the location of the country's atomic research institute, where the element was first made.

SEABORGIUM (106)

Named after Glenn Seaborg, who became the first person to have an element named after them while they were still alive.

DARMASTADTIUM (110)

This element is thought to be a metal with similar properties to platinum, but its most stable isotope has a half-life of just ten seconds.

ROENTGENIUM (111)

Named after the discoverer of X-rays, Wilhelm Röntgen, this element is predicted to be a noble metal, like silver and gold.

COPERNICIUM (112)

Named for Nicolaus Copernicus, who showed that Earth orbited the Sun, not the other way round. Copernicium is predicted to be a metal but also a gas in standard conditions.

LIVERMORIUM (116)

All of the half-lives of this highly radioactive element are measured in milliseconds.

TENNESSINE (117)

The heaviest halogen, this member of Group 7 is predicted to be a metal, with physical properties similar to lead.

OGANESSON (118)

A new noble gas, this element is named after the Russian researcher Yuri Oganessian, the only person still alive to have an element named after him. A single isotope of oganesson has a half-life of 0.7 milliseconds.

THE FUTURE?

If more superheavy elements are produced in the future, they will begin an eighth period. The eminent American particle physicist Richard Feynman predicted that the largest atom possible would be element 137 (dubbed 'feynmanium'). Above that size, neutrons would spontaneously collapse. Other scientists disagree. We will have to wait and see.

GLOSSARY

alloy A mixture of metals.

alpha particle A particle produced by certain radioactive decays. The particle is made of two protons and two neutrons, and has a charge of +2.

anion A negatively charged ion.

atom The smallest unit of an element. The atom can be simplified into smaller particles – protons, neutrons and electrons.

atomic number The number of protons in an atom. The atoms of a specific element always have the same atomic number.

atomic mass A measure of the number of particles in an atom's nucleus.

catalyst A substance that speeds up the rate of a chemical reaction between two or more other substances. However, the catalyst is not used up in the reaction.

cation A positively charged ion.

charge An electrical quality of subatomic particles, which is conferred on larger objects; objects with opposite charges attract each other, while those with similar ones repel.

compound A substance made by the atoms of two or more different elements forming chemical bonds.

decay When an unstable atom breaks down, transforming into an atom of another element

electron A negatively charged particle found in atoms.

fission When an atom splits into two roughly equally sized atoms.

fusion When two small atoms fuse to make one larger one.

group The name used for the columns of the periodic table; members of a group share many properties.

half-life The time it takes for a sample of unstable, radioactive atoms to decrease in size by half; this is used as

a measure of a substance's instability.

ion An atom that has lost or gained electrons to become a charged object.

isomer A molecule that contains the same number and type of atoms, but arranges them in different ways.

isotope A version of an atom that has a different number of neutrons in its nucleus. Elements all exist in several isotopes.

mole The standard unit of quantity; moles are used to measure the number of atoms and molecules.

molecule The smallest unit of a compound.

neutron A particle found in the nucleus of all atoms, except the most common isotope of hydrogen; neutrons do not have a charge.

nucleus The core of an atom that contains almost all of its mass.

orbital A space around an atom's nucleus where

electrons are located.

period The name used for the rows of the periodic table, so called because the elements included show a regular, or periodic, pattern of changes in their chemical properties.

positron A positively charged antimatter version of an electron produced during certain types of radioactive decay.

proton A positively charged particle that is located in the nucleus of every atom. Every element has a unique number of protons in their nuclei.

quark A subatomic particle, of similar size to an electron, that makes up protons, neutrons and more exotic particles.

radioactivity A behaviour of unstable elements, where the forces holding protons and neutrons together inside the nucleus are not able to hold it together so it eventually decays, or releases mass and energy.

radius The distance from the centre of a circle or sphere to the outside.

series A set of elements in the periodic table that do not follow the normal configuration of electrons in their atoms.

smelting A chemical process used to make pure metal from an ore.

superconductor A substance that offers no resistance to an electrical current.

transition metal A member of the largest series in the periodic table; they form the middle section, or transition, of the table.

valence A measure of how many bonds an atom can form with other atoms.

wavelength A measure of the distance between the peak of one wave with the one behind; the wavelength of a light wave is a direct measure of how much energy it contains.

INDEX

ACKNOWLEDGMENTS

All main illustrations © Andi Best
Except photographic images on the following pages:

4-5 © arleksey/Shutterstock.com; 6-7 © tj-rabbit/Shutterstock.com; 14 © 3D Vector/Shutterstock.com; 15 © By TuiPhotoEngineer/Shutterstock.com; 21A © By koya979/Shutterstock.com, 21B © By JIANG HONGYAN/Shutterstock.com, 21C © By Dim Dimich/Shutterstock.com, 21D © By Dim Dimich/Shutterstock.com; 23A © By eAlisa/Shutterstock.com, 23B © By Chansom Pantip/Shutterstock.com, 23C © By Rob Wilson/Shutterstock.com, 23D © By Petr Novotny/Shutterstock.com; 27A © By koya979/Shutterstock.com, 27B © By mountainpix/Shutterstock.com, 27C © By Denys Dolnikov/Shutterstock.com, 27D © By Artography/Shutterstock.com, 29A © By rCarner/Shutterstock.com, 29B © By LukaKikina/Shutterstock.com, 29C © By TuiPhotoEngineer/Shutterstock.com, 29D © By Chones/Shutterstock.com, 29E © By Oldrich/Shutterstock.com; 31A © By RACOBOVT/Shutterstock.com, 31B © By rosesmith/Shutterstock.com, 31C © By USJ/Shutterstock.com, 31D © By Africa Studio/Shutterstock.com, 31E © By Aleksey Klints/Shutterstock.com; 36 © By Vladystock/Shutterstock.com; 48A © By Thammasak Lek/Shutterstock.com, 48B © By Volodymyr Goinyk/Shutterstock.com, 48C © By boykung/Shutterstock.com, 48D © By rCarner/Shutterstock.com, 48E © By Oleksandr Lysenko/Shutterstock.com, 48F © By Jojje/Shutterstock.com; 49 © By cigdem/Shutterstock.com; 65A © By boykung/Shutterstock.com, 65B © By Mivr/Shutterstock.com, 65C © By Cagla Acikgoz/Shutterstock.com, 65D © By Fribus Mara/Shutterstock.com, 65E © By Aleksandr Pobedimskiy/Shutterstock.com, 65F © By Coldmoon Photoproject/Shutterstock.com, 65G © By dorky/Shutterstock.com, 65H © By Jiri Vaclavek/Shutterstock.com, 65I © By Aleksandr Pobedimskiy/Shutterstock.com, 65J © By Bramthestocker/Shutterstock.com; 90A © By boykung/Shutterstock.com, 90B © By gresei/Shutterstock.com, 90C © By joingate/Shutterstock.com, 90D © By Fotofermer/Shutterstock.com; 91A © By Peangdao/Shutterstock.com, 91B © By ifong/Shutterstock.com, 91C © By Subject Photo/Shutterstock.com; 102A © By Ensuper/Shutterstock.com, 102B © By decade3d/Shutterstock.com, 102C © By Ian 2010/Shutterstock.com, 102D © By Valentyn Volkov/Shutterstock.com, 102E © By Janthiwa Sutthiboriban/Shutterstock.com, 102F © By Iraidka/Shutterstock.com, 102G © By Juris Surainis/Shutterstock.com; 103A © By Karin Hildebrand Lau/Shutterstock.com, 103B © By totojang1977/Shutterstock.com, 103C © By oksana20/Shutterstock.com, 103D © By AlenKadr/Shutterstock.com, 103E © By manusy/Shutterstock.com; 111A © By masa44/Shutterstock.com, 111B © By Alexey Boldin/Shutterstock.com, 111C © By KREML/Shutterstock.com, 113A © By Guillermo Pis Gonzalez/Shutterstock.com, 113B © By M. Unal Ozmen/Shutterstock.com, 113C © By cobalt88/Shutterstock.com, 113D © By Photo Love/Shutterstock.com, 113E © By Fotokostic/Shutterstock.com; 114A © By azure1/Shutterstock.com, 114B © By VanderWolf Images/Shutterstock.com, 114C © By Michal Sanca/Shutterstock.com, 114D © By pattang/Shutterstock.com, 114E © By Nyvlt-art/Shutterstock.com; 115A © By Vereshchagin Dmitry/Shutterstock.com, 115B © By Philmoto/Shutterstock.com, 115C © By Rawpixel.com/Shutterstock.com; 116A © By Andrii Symonenko/Shutterstock.com, 116B © By Pablo del Rio Sotelo/Shutterstock.com, 116C © By pattang/Shutterstock.com, 116D © By Madlen/Shutterstock.com, 116E © By Peter Sobolev/Shutterstock.com, 117A © By Peter Sobolev/Shutterstock.com, 117B © By Adam Vilimek/Shutterstock.com; 120A © By Somchai Som/Shutterstock.com, 120B © By Merydolla/Shutterstock.com; 121A © By WhiteBarbie/Shutterstock.com; 122 © By Dennis Owusu-Ansah/Shutterstock.com; 123 © By Zanna Art/Shutterstock.com; 124A © By xxxx/Shutterstock.com, 124B © xxxx/Shutterstock.com; 125A © By Sergiy Kuzmin/Shutterstock.com, 125B © By Vlad Kochelaevskiy/Shutterstock.com, 125C © By slhy/Shutterstock.com, 125D © By Rawpixel.com/Shutterstock.com; 126A © By Gloriole/Shutterstock.com, 126B © By wk1003mike/Shutterstock.com, 126C © By xxxx/Shutterstock.com, 126D © By Oleksandr Rybitskiy/Shutterstock.com; 127A © By donatas1205/Shutterstock.com, 127B © By Lorant Matyas/Shutterstock.com, 127C © By koosen/Shutterstock.com; 131A © By Fotovika/Shutterstock.com, 131B © By Luisa Puccini/Shutterstock.com, 131C © By Vachagan Malkhasyan/Shutterstock.com, 131D © By Garsya/Shutterstock.com; 132A © By tale/Shutterstock.com, 132B © By Jeff Whyte/Shutterstock.com, 132C © By Scanrail/Shutterstock.com, 132D © By Garsya/Shutterstock.com; 133A © By Medical Art Inc/Shutterstock.com, 133B © By Hein Nouwens/Shutterstock.com, 133C © By bergamont/Shutterstock.com, 133D © By Yevhenii Popov/Shutterstock.com, 133E © By Brilliance stock/Shutterstock.com, 133F © By Kalin Eftimov/Shutterstock.com, 133G © By Abramova Elena/Shutterstock.com, 133H © By JIANG HONGYAN/Shutterstock.com, 133I © By Superheang168/Shutterstock.com, 133J © By AlenKadr/Shutterstock.com; 134A © By noreefly/Shutterstock.com, 134B © By ifong/Shutterstock.com; 137A © By Panos Karas/Shutterstock.com, 137B © By Nikandphoto/Shutterstock.com, 137C © By l000s_pixels/Shutterstock.com; 138A © By Volodymyr Krasyuk/Shutterstock.com, 138B © By Taigi/Shutterstock.com, 138C © By Sashkin/Shutterstock.com; 139A © By Oleksandr Kostiuchenko/Shutterstock.com, 139B © By Alexey Boldin/Shutterstock.com, 139C © By horiyan/Shutterstock.com, 139D © By Rawpixel.com/Shutterstock.com, 139E © By wk1003mike/Shutterstock.com, 139F © By s-ts/Shutterstock.com; 144A © By Planner/Shutterstock.com, 144B © By Joshua Resnick/Shutterstock.com, 144C © By Lorant Matyas/Shutterstock.com, 144D © By Madlen/Shutterstock.com, 144E © By Constantine Pankin/Shutterstock.com; 145A © By Ivelin Radkov/Shutterstock.com, 145B © By Elenarts/Shutterstock.com, 145C © By ar3ding/Shutterstock.com, 145D © By Richard Peterson/Shutterstock.com; 146A © By NikoNomad/Shutterstock.com, 146B © By gritsalak karalak/Shutterstock.com, 146C © By Konstantin Faraktinov/Shutterstock.com; 148 © By Yurchyks/Shutterstock.com; 149A © By Dmitrydesign/Shutterstock.com, 149B © By Oliver Hoffmann/Shutterstock.com, 149C © By Mariyana M/Shutterstock.com, 149D © By drawhunter/Shutterstock.com, 149E © By elnavegante/Shutterstock.com; 150A © By Ninell/Shutterstock.com, 150B © By horiyan/Shutterstock.com, 150C © By stockphoto mania/Shutterstock.com, 150D © By Oleksandr Kostiuchenko/Shutterstock.com, 150E © By mtlapcevic/Shutterstock.com; 151A © By Peangdao/Shutterstock.com, 151B © By ifong/Shutterstock.com, 151C © By Subject Photo/Shutterstock.com, 151D © By MOAimage/Shutterstock.com, 151E © By Kichigin/Shutterstock.com; 152A © By Jojje/Shutterstock.com, 152B © By KREML/Shutterstock.com, 152C © By Kulakov Yuri/Shutterstock.com, 152D © By A. L. Spangler/Shutterstock.com; 154A © By Aaron Amat/Shutterstock.com, 154B © By Marco Vittur/Shutterstock.com, 154C © By James Steidl/Shutterstock.com, 154D © By RACOBOVT/Shutterstock.com, 154E © By Vereshchagin Dmitry/Shutterstock.com; 159 © By Triff/Shutterstock.com; 160A © By Mr. SUTTIPON YAKHAM/Shutterstock.com, 160B © By Vereshchagin Dmitry/Shutterstock.com, 160C © By Jojje/Shutterstock.com, 160D © By Pavel Chagochkin/Shutterstock.com, 160E © By Vachagan Malkhasyan/Shutterstock.com, 160F © By Nerthuz/Shutterstock.com, 160G © By Rawpixel.com/Shutterstock.com; 162 © By Johannes Kornelius/Shutterstock.com; 167 © By Kvadrat/Shutterstock.com; 168A © By Sergey Peterman/Shutterstock.com, 168B © By Sofiaworld/Shutterstock.com, 168C © By VFilimonov/Shutterstock.com, 168D © By Sementer/Shutterstock.com; 169A © By decade3d - anatomy online/Shutterstock.com, 169B © By Panda Vector/Shutterstock.com; 170 © By Chones/Shutterstock.com; 173A © By VFilimonov/Shutterstock.com, 173B © By Cronislaw/Shutterstock.com, 173C © By ARTEKI/Shutterstock.com; 175A © By Accurate shot/Shutterstock.com, 175B © By Elnur/Shutterstock.com, 175C © By Fruit Cocktail Creative/Shutterstock.com; 176A © By Mauro Rodrigues/Shutterstock.com, 176B © By Jiri Hera/Shutterstock.com, 176C © By Andrey Lobachev/Shutterstock.com; 177A © By cigdem/Shutterstock.com, 177B © By NPeter/Shutterstock.com; 178 © By Frederic Legrand - COMEO/Shutterstock.com; 179A © By cobalt88/Shutterstock.com, 179B © By Yuliyan Velchev/Shutterstock.com; 180 © Abert/Shutterstock.com; 181 © cobalt88/Shutterstock.com; 183 © James Steidl/Shutterstock.com; 184 © Palomba/Shutterstock.com; 185 © eAlisa/Shutterstock.com; 189 © Nicolas Primola/Shutterstock.com; 191A © Somchai Som/Shutterstock.com, 191B © MIGUEL GARCIA SAAVEDRA/Shutterstock.com; 193A © Beautyimage/Shutterstock.com, 193B © Somchai Som/Shutterstock.com, 193C © Andrey Burmakin/Shutterstock.com; 193D © Pan Xunbin/Shutterstock.com; 194A © Johan Swanepoel/Shutterstock.com, 194B © Ozja/Shutterstock.com, 194C © Hedzun Vasyl/Shutterstock.com; 197 © stockphoto mania/Shutterstock.com; 199A © Vachagan Malkhasyan/Shutterstock.com, 199B © Gino Santa Maria/Shutterstock.com; 200A © Carlos Romero/Shutterstock.com, 200B © Lukasz Grudzien/Shutterstock.com, 200C © Volodymyr Krasyuk/Shutterstock.com, 200D © Petr Salinger/Shutterstock.com; 201D © www.auditionsfree.com; 202A © Susan Schmitz/Shutterstock.com, 202B © Kazakov Maksim/Shutterstock.com, 202C © Jojje/Shutterstock.com, 202D© Libor Fousek/Shutterstock.com, 202E © Aleksey Stemmer/Shutterstock.com, 202F © SOMMAI/Shutterstock.com, 202G © serg_dibrova/Shutterstock.com, 202H © Aleksey Klints/Shutterstock.com; 205A © Sashkin/Shutterstock.com, 205B © Cozy nook/Shutterstock.com; 206A © pattang/Shutterstock.com, 206B © TairA/Shutterstock.com, 206C © RACOBOVT/Shutterstock.com; 207A © Kostsov/Shutterstock.com; 212A © NPeter/Shutterstock.com, 212B © Vadim Sadovski/Shutterstock.com, 212C © Vadim Sadovski/Shutterstock.com, 212D © NASA images/Shutterstock.com; 213 © KREML/Shutterstock.com; 215A © u3d/Shutterstock.com, 215B © ESA/ATG medialab, 215C © Jaroslav Moravcik/Shutterstock.com, 215D © tsuneomp/Shutterstock.com; 217 © stockphoto mania/Shutterstock.com